KU-760-037

Scientific Writing for Young Astronomers
Part 2

EAS Publications Series, Volume 50, 2011

Scientific Writing for Young Astronomers

A Collection of Papers on Scientific Writing

Part 2

Edited by: *Christiaan Sterken*

EDP
SCIENCES

17 avenue du Hoggar, PA de Courtabœuf, B.P. 112, 91944 Les Ulis cedex A, France

First pages of all issues in the series and full-text articles
in PDF format are available to registered users at:
http://www.eas-journal.org

Cover Figure

The tapestry *Astronomie* depicting the Muse *Astronomia*.
Linen, wool and silk, 240 × 340 cm, made in Paris, *circa* 1500–1510.
Photographed by Swedish photographer Håkan Berg.
ⓒ Reproduced with permission of the *Röhsska Museum of Fashion, Design and Decorative Arts* (*Röhsska Museet för Konsthantverk och Design*), Göteborg, Sweden.

Indexed in: ADS, Current Contents Proceedings – Engineering & Physical Sciences, ISTP®/ISI Proceedings, ISTP/ISI CDROM Proceedings.

ISBN 978-2-7598-0639-3   EDP Sciences Les Ulis
ISSN 1633-4760
e-ISSN 1638-1963

Table of Contents

Expanded Table of Contents

Preface

The *Astronomy & Astrophysics* (A&A) Board of Directors and EDP Sciences, the Publisher of A&A, decided in 2007 to organize a School on the various aspects of scientific writing and publishing, with two aims: (a) to teach young PhD students how to express their scientific thoughts and results through adequate and efficient written communication, and (b) to discuss the operation of A&A as an example of an international peer-reviewed journal in astronomy. To this end, a first *Scientific Writing for Young Astronomers* (SWYA) School was organized in 2008, and a second one in 2009. The Schools were financed by A&A and by EDP Sciences, who also provided all logistic support.

This book, together with Volume 49 of the EAS Publications Series, is the outcome of these Schools. Whereas Volume 49 presents guidelines and examples for publishing in academic journals (like A&A), the three papers in this Volume are aimed at supplying guidelines to PhD students and postdoctoral fellows to help them compose scientific papers for different forums (journals, proceedings, thesis manuscripts, etc.). These papers address the writing process, graphics, ethics, and bibliometry, and cover the information that was presented in about a dozen lectures. These chapters feature several examples and case studies in astronomical context, in addition to many examples from scientific enquiry in a broader sense – even from art and from business life.

Paper I copes with the preparation of manuscripts, with the handling of copyrights and permissions to reproduce, with communicating with editors and referees, and with avoiding common errors. More than two dozen FAQs (on authorship, on refereeing, on revising multi-authored papers, etc.) are answered.

Paper II is entirely dedicated to communication with graphics, *i.e.*, to all facets of visual communication by way of images, graphs, diagrams and tabular material. Design types of graphs are explicated, as well as the major components of graphical images. The basic features of computer graphics are explained, as well as concepts of color models and of color spaces (with emphasis on color graphics for viewers suffering from color-vision deficiencies). Special attention is given to the verity of graphical content, and to misrepresentations and errors in graphics and in associated basic statistics. Dangers of dot joining and curve fitting are discussed, with emphasis on the perception of linearity, the issue of nonsense correlations, and the handling of outliers. The remainder of the chapter illustrates the distinction between data, fits and models.

The main theme of Paper III is truthful communication of scientific results, and hinges on the pillar that every scientist daily comes across: **ethics** – involving two major aspects of research, *viz.*, the measurement of scientific value, and the enforcement of proper conduct in research and in scientific writing. The following bibliometric parameters are explained: the journal impact factor, the journal cited half-life, and the journal immediacy index, as well as paper counts, citation rates, citation index and the Hirsch index. These bibliometric indices and indicators are illustrated with real examples derived from bibliometric analyses of the astronomical literature. The biases of bibliometric indices, and the use and abuse of bibliometrics are worked out. Moreover, suggestions for remediating the present system of bibliometric measurement and evaluation are made. Scientific misconduct in the broadest sense is discussed by category: researcher misconduct, author misconduct, referee and grant-reviewer misconduct. But also publisher misconduct, editorial misconduct and mismanagement, and research supervisor misbehavior are dealt with. The overall signatures of scientific misconduct are focussed on, as well as the causes and the cures. This is followed by a Section devoted to whistleblowing. This chapter illustrates the complex endeavor that education in science really is, and also comprises passages that directly deal with the hopes and concerns of our students, either expressed during and after the lectures, or through private contacts later on.

The best way to use this book is to read it twice: a first reading for skimming its content, and then a deeper reading of the passages that are relevant to what the reader wants to know. Or use this book as a reference, and read the Sections corresponding to items listed in the index or in the expanded Table of Contents.

The opinions expressed in these papers are those of the author, and those of the authors of the cited papers, and do not necessarily reflect the views of the Editors of A&A. None of these papers were submitted for language editing to a professional language editor.

The Schools took place in Hotel and Conference Center Aazaert in the city of Blankenberge on the North Sea coast of Belgium. The meetings were very rewarding, and happened in a very pleasant atmosphere, thanks to the dedication and professionalism of the complete staff: from the barkeeper to the waiters, and from the administrative staff to the chefs in the kitchen – they all contributed greatly in making us feel at home, and in giving us an unforgettable time. In the name of all participants, I express deep appreciation to Mr. Peter Kamoen, Manager of Hotel Aazaert, and to his entire staff.

The Manager of the Schools (on behalf of EDP) was Mrs. Isabelle Turpin, assisted by Mrs. Aurélie Chastaingt. They did a professional job preparing each event, and their presence during the complete duration of each School was a blessing for anyone who needed help. In the name of all lecturers and students, I extend a warm thanks to Isabelle and to Aurélie.

Dr. Agnès Henri, Mrs. Sophie Hosotte, Mrs. Isabelle Houlbert and Mrs. Marie-Louise Chaix (EDP) are given thanks for their most adequate communication and helpful support during the preparation of this book.

My special thanks go to those students who – either during or after the lectures, or through subsequent private contacts – shared with me their hopes and concerns.

I am very much indebted to Dr. Hilmar W. Duerbeck, Dr. Chris Koen, and Dr. Laszlo Szabados for deep proofreading of, and criticism, discussion and advice on the three papers included in this book.

I acknowledge financial support from the Vrije Universiteit Brussel, and I express gratitude to my Employer, the Belgian Fund for Scientific Research (FWO), for supporting several research projects that, over the years, gave me the opportunity to acquire the specialized knowledge presented in this book.

Christiaan Sterken
Editor
University of Brussels, Belgium

DOI: 10.1051/eas/1150000

Acronyms

AAS	American Astronomical Society
AD	Anno Domini
ADS	NASA Astrophysics Data System
AJ	Astronomical Journal
ASCII	American Standard Code for Information Interchange
ASP	Astronomical Society of the Pacific
A&A	Astronomy & Astrophysics
ApJ	Astrophysical Journal
ApJS	Astrophysical Journal Supplement Series
BC	Before Christ
BCE	Before the Common Era
BMI	Body Mass Index
BMJ	The British Medical Journal
BMP	Windows bitmap format
Caltech	California Institute of Technology
CCC	Copyright Clearance Center
CCD	Charge Coupled Device
CDS	Centre de Données astronomiques de Strasbourg
CIE	Commission Internationale de l'Éclairage
CMMI	Computer Modern Math Italic
COPE	Committee on Publication Ethics
CRT	Cathode Ray Tube
CV	Curriculum Vitae
DOI	Digital Object Identifier
dpi	dots per inch
DSS	Digital Sky Survey
EDP	EDP Sciences
EPS	Encapsulated Postscript
ESA	European Space Agency
ESO	European Southern Observatory
FAQ	Frequently Asked Questions
FFP	Fabrication, Falsification and Plagiarism
FITS	Flexible Image Transport System
GCPD	General Catalogue of Photometric Data
GIF	Graphics Interchange Format
HJD	Heliocentric Julian Date
HST	Hubble Space Telescope
IAU	International Astronomical Union
IBVS	Information Bulletin on Variable Stars
ICMJE	International Committee of Medical Journal Editors

IF	Journal Impact Factor
IMRAD	Introduction, Methods, Results and Discussion
IQ	Intelligence Quotient
IRAS	Infrared Astronomical Satellite
ISI	Institute for Scientific Information
ISO	Infrared Space Observatory
JCR	Journal Citation Report
JPEG	Joint Photographic Experts Group
LBV	Luminous Blue Variable
LISA	Library and Information Services in Astronomy
LMC	Large Magellanic Cloud
lpi	lines per inch
MJD	Modified Julian Date
MMS	Manuscript Management System
MNRAS	Monthly Notices of the Royal Astronomical Society
NASA	National Aeronautics and Space Administration
OA	Open Access
OCR	Optical Character Recognition
$O-C$	Observed minus Calculated
ORCID	Open Researcher & Contributor ID
PASP	Publications of the Astronomical Society of the Pacific
PDF	Portable Document Format
PICT	Apple Picture Format
PMT	Photomultiplier Tube
PNG	Portable Network Graphics
ppi	points per inch
pt	printer's point
QPO	Quasi Periodic Oscillation
RAS	Royal Astronomical Society
SAO	Smithsonian Astrophysical Observatory
SDSS	Sloan Digital Sky Survey
SI	International System of Units
SIMBAD	Set of Identifications, Measurements, and Bibliography for Astronomical Data
SPIRES-HEP	High Energy Physics database, Stanford
S/N	Signal to noise ratio
SWYA	Scientific Writing for Young Astronomers
TAI	International Atomic Time
TIFF	Tagged Image File Format
URL	Uniform Resource Locator
USNO	United States Naval Observatory
UT	Universal Time
UTC	Coordinated Universal Time
VLT	Very Large Telescope
WAME	World Association of Medical Editors
WET	Whole Earth Telescope
WoS	Web of Science
ZAMS	Zero-age main sequence

About the Authors

Christiaan Sterken is Research Director of the *Belgian Fund for Scientific Research* (Fonds Wetenschappelijk Onderzoek), and works at the Department of Physics at the University of Brussels (Vrije Universiteit Brussel). His principal field of research is the photometry of variable stars (luminous blue variables, massive binaries, pulsating main sequence stars, and cataclysmic variables), and comets. He also teaches courses in observational astronomy, and on the history of natural sciences at the University of Brussels.

Scientific Writing for Young Astronomers
C. Sterken (ed)
EAS Publications Series, **50** (2011) 1–63
www.eas-journal.org

WRITING A SCIENTIFIC PAPER I.
THE WRITING PROCESS

C. Sterken[1]

Abstract. Based on my direct personal experience as an author, referee, editor, and PhD thesis supervisor, I present in this paper some advice and guidelines on writing scientific papers. This article copes with the preparation of manuscripts, with the handling of copyrights and permissions to reproduce, with communicating with editors and referees, and with avoiding common errors. More than two dozen FAQs are dealt with, among others: who should be coauthor, what is honorary authorship, how to settle the order of author names, may the list of authors change during the review process, who submits the manuscript to the Editor, must one wait for coauthors' consent to submit, when to submit a manuscript to `astro-ph`, how to communicate with the referee, with whom to share a referee report, can students be referee, how to cite and quote a straying paper, how to cope with writer's block, should others be allowed to completely rewrite a manuscript, how to review a paper, how to revise a multi-author paper, when to ask for another referee, why do Editors edit, and why do referees review?

Preamble

This paper is the first of three expanded sequels to *Advice on Writing a Scientific Paper*[1], Sterken (2006), exploring key aspects of the scientific writing process. In particular, advice is given on how to prepare manuscripts, handle copyrights and permissions to reproduce, how to communicate with Editors and referees, how to avoid common errors, what are the requirements for demanding authorship, and which guidelines to follow when determining a fair authorship sequence. The second paper handles graphics (images, graphs and Tables), and the third paper deals with ethics in scientific communication.

As these papers (hereafter referred to as Paper I, Paper II and Paper III, respectively) deal with tutoring on writing and publishing, the reader will notice some overlap

---

[1]Complete text freely available at `http://www.aspbooks.org/publications/349-0445.pdf`

among these three articles, and also between these papers and the other papers in these Proceedings. Such repetition is unavoidable, since every lecturer's approach is from a different viewpoint (Editor, referee, author, thesis supervisor, ...), but the redundancy is also included on purpose since repetition is an innate element of the practice of teaching. Papers I–III also present ample advice on research practice through teaching by example.

These papers are written in the first and second person singular – a not so common approach in scientific communication that commonly utilizes *passive voice* – but since they deal with somewhat personal advice based on hard-won experience, this approach is much more direct and persuasive than when speaking in the third person. The reader is thus addressed in the second person, the target audience being the young astronomer at the beginning of her or his life as an author (though senior researchers may occasionally take profit of the advice presented here). As the SCIENTIFIC WRITING FOR YOUNG ASTRONOMERS (SWYA) schools were organized on behalf of the journal *Astronomy & Astrophysics* (A&A), some of the practical advice given here is tuned for publishing in this specific journal.

In these three papers, I proceed on the premise that the reader is strongly committed to the core business of science: *verity* and *truth*, even more so to the *truthful communication* of scientific findings (see also Sect. 5.17). Paper II, and the Epilog of Paper III, address an additional element of science and scientific communication: *beauty* and *aesthetics*.

1   Introduction

The structure of this paper is as follows. The next three Sections deal with the question why we write and why we publish, and what the principal differences are between the spoken word (lecture) and written (published) communication. Then comes a Section describing the different categories of research papers, followed by three Sections dealing with language, and with the reading and the writing process. A complete Section is devoted to the structure of a scientific paper. Finally, a selection of FAQs on practical matters is presented.

2   Why do scientists write?

We are so used to see facts and theories as the only building blocks of science, that we often forget that there is much more, as is so well conveyed by Shermer (2007):

> TO BE OF TRUE SERVICE TO HUMANITY, SCIENCE MUST BE AN EXQUISITE BLEND OF DATA, THEORY AND NARRATIVE.

Data and theory are complemented and completed by *narration*. Any narration or storytelling consists of three elements: *i)* the **content** of the message or story, *ii)* the **form** of the story, and *iii)* the **direction** of the communication (author to reader, author to Editor, author to referee). Moreover, the writing phase of a research project not seldomly modifies the research conclusions, as is testified by Perkowitz (1989):

> I HAVE LEARNED THAT WHEN I WRITE A RESEARCH PAPER I DO FAR MORE THAN SUM-MARIZE CONCLUSIONS ALREADY NEATLY STORED IN MY MIND. RATHER, THE WRITING

PROCESS IS WHERE I CARRY OUT THE FINAL COMPREHENSION, ANALYSIS, AND SYNTHE-
SIS OF MY RESULTS.

Thus, without publication, research results remain incomplete. This cognitive process is known as *advancing insight*: ideas that arise as the work evolves. Umberto Eco also says[2]

WRITING IS AN ACT OF COMMUNICATION, IT'S AN ACT OF LOVE, IT'S SOMETHING YOU DO IN ORDER TO BE UNDERSTOOD.

Perkowitz's and Eco's statements underline that it is very important to learn how to properly produce narrative (*i.e.*, write texts, or make good graphics), but also how to consume narrative (*i.e.*, how to read papers, and how to understand referee reports). The main aim of writing is communication: above all, **we write to be read** – and not to be cited as a first purpose.

Before embarking on the writing of papers, it is useful to first deal with the question of why we publish, then to concentrate on the crucial difference between the printed word *versus* the oral message, and finally to explain the reading process.

3   Why do scientists publish?

Scientific research is a collective undertaking, and involves the distribution of knowledge, and the sharing of results. This dissemination of knowledge, using established rules and standards of scientific reporting, is the prime reason for publishing. Evidently, we do not always and only publish with this motive in mind; in real practice there are several gradations in the incentive for writing. The following short paragraphs list, in order of diminishing importance, some motivations for publishing.

1. BECAUSE I WANT TO REPORT NEW SCIENTIFIC RESULTS AND GET THE CREDIT. Besides the wish and need to communicate results, it is very important to publish as soon as new results and discoveries are considered solid. Credit is usually (but not always) attributed to the person who first makes a result public. Early publication allows another necessary process to commence: *experimental verification* of the observations or theories – and, perhaps, *refutation*.

2. BECAUSE I NEED A JOB, A PROMOTION, OR A GRANT. Any successful application for a job or a research grant requires proof of scientific creativity and productivity. This evidence indirectly emanates from the publication record. A colleague once wrote me that he sought a contract renewal and, therefore, needed to produce *"a paper, quick and dirty"*. The resulting quick paper may have been effective in his pursuit of a job, but needless to say, the paper cannot have been on a high standard of creativity if the job hunt was the only motive for writing it.

3. BECAUSE I WANT TO ACHIEVE SOCIAL CLIMBING BY BEING VISIBLE IN ADS[3]. This ambition to gain esteem by fellow scientists is somehow related to the previous one, in fact to the

---

[2]Quoted by Mahtot Teka, Time Magazine, 10 December 2007.
[3]The NASA Astrophysics Data System, http://www.adsabs.harvard.edu/

complete mechanism of "measuring" scientific stature by "counting" a person's papers and citations through bibliometric tools like ADS (see Sect. 5.2 in Paper III for a more profound discussion).

4. BECAUSE I AM TRAVELING, AND PAPERING IS THE ONLY WAY TO COVER MY TRAVEL COSTS. This is unfortunate but realistic: many conference attendees get their expenses covered only if they present a paper (in practice, a poster is displayed most of the time).

The first motive for publishing is the most earnest: announce new results, describe new findings, or report significant errors in published works. I could not say it clearer than the Chilean artist Violeta Parra (1917–1967) does in *Yo canto a la diferencia*:[4]

> ...I DO NOT PLAY THE GUITAR FOR APPLAUSE. I SING OF THE DIFFERENCE BETWEEN WHAT IS TRUE AND WHAT IS FALSE. OTHERWISE, I DO NOT SING.

It all boils down to the amount of new-information content of your paper: if, for example, the information content of a poster paper is insignificant – either because of poor content, or because the poster is just a duplicate of another paper (going to be) published elsewhere – then the poster paper simply is irrelevant, and should not be published in the first place.

4   Printed paper *versus* oral communication

There are fundamental differences between oral and literate modes of communication – between a talk and a written paper – such as:

- the printed message is not instantaneous: whereas the speaking voice is a direct address to an audience, the printing and distribution process takes months of time;

- speaking is continuous communication, the printed message is discontinuous;

- printed text does not render any intentional or inadvertent body language – bodily movement, posture or facial expression, but also vocal stress and changes in pitch – that underlines and even amplifies the message;[5]

- a printed message does not provide any non-verbal "leakage", *i.e.*, hesitations and signs of poor preparation – though printed text may carry equivalent signs of sloppiness, such as typographical errors;

- a written paper calls for careful wording in order not to be misunderstood, or in order to avoid being cited out of context, (see also Sect. 11.9);

- a printed text allows the author to present a very different degree of detail in the description of the research procedures and results;

---

[4] *...no tomo la guitarra por conseguir un aplauso. Yo canto a la diferencia que hay de lo cierto a lo falso. De lo contrario, no canto.*

[5] There are many languages of gesture, and facial expressions are not always intelligible to those raised in another culture, see Thomas (1991).

- because the spoken word is straight communication, it provides for instantaneous feedback;

- whereas a confusing idea – whether ambiguously described or inherently complex – may get across the attentive reader's mind after a second reading of a same phrase, the time frame of the spoken word does not allow such a second pass: even simple messages do not always penetrate the receiver's mind in real time, because attention occasionally drops off to the level that *you may have heard the message, though without really listening*[6] *to it*;

- published phrases remain forever, spoken words often persist only for a very short time;

- whereas sitting through a bad lecture may possibly yield a minor bonus (*e.g.*, when the lecturer uses well-designed slides and attractive animations), a poorly written paper has no value whatsoever, since such papers *always* convey the message that *"what is poorly written is also of doubtful scientific value"* (Stein 2006);

- a lecture – an invited lecture above all – is a social event, with a direct relationship between speaker and audience, and with considerable emphasis on the pedagogic form. Reading a written text is a more solitary experience.

A published paper is quite often the written record of a talk, like some of the papers in these Proceedings. Whereas one just cannot publish a *verbatim* record of a lecture, the written article normally corresponds to the talk (at least in content). In this set of three papers, however, I deliberately deviate from this principle, because the matter covered was spread out over almost a dozen lectures in both SWYA schools.

5   Categories of scientific papers

There are many categories of scientific papers. This Section describes the most common types, as roughly ranked by importance (see also the A&A website `www.aanda.org`).

5.1   *Regular research paper in a refereed journal*

Submission of a paper to a refereed journal (like A&A) mostly implies that the paper is the result of original research not previously published, nor submitted for publication elsewhere, nor considered for later reproduction without the consent of the copyright holder. Note that some refereed journals will not even submit to the referee a paper that is "too light". Bertout & Schneider (2004), for instance, give a typical example of the type of topic that they deem unsuitable for submission to A&A: *"... an unremarkable binary together with a standard interpretation of its light curve"*. It is thus advisable to check the editorial policies of a journal, and verify what types of papers are published by that journal.

---

[6]The verb *to hear* means to perceive or apprehend by the ear, the verb *to listen* means to hear something with thoughtful attention (*Merriam–Webster Collegiate Dictionary*).

Do not forget that papers that are not refereed are discounted for career promotions and job applications. Most refereed journals require that manuscripts be submitted in the so-called referee format (double spacing: full page instead of two-column format), so that the referee can make annotations.

5.2 Letter to the Editor

A *Letter* is a research paper (or a short opinion paper) that requires rapid publication. Some types of Letters are esteemed higher than regular refereed papers, others have a much lower esteem. A&A publishes Letters to communicate important new results that require rapid publication. They are restricted in length to four printed pages, and are usually published within 4–8 weeks of acceptance (see the paper by Claude Bertout in Part 1 of these Proceedings). Whether a manuscript is accepted as a Letter, or is diverted for publication as a regular paper, is the decision of the Editor, though the opinion of the referee not seldomly is the determining factor.

5.3 Research Note

A&A accepts *Research Notes*, *i.e.*, short papers that contain either new results as an extension of work reported in a previous paper, or limited observations not urgent enough to be published as a Letter, or useful calculations that have no definite immediate astrophysical applications. A Research Note can also present a preliminary analysis of a large dataset deemed important for timely communication to the readership of the journal.

5.4 Review paper

A review paper is a wide retrospective survey of a specific field, including a critical evaluation of the scientific subdiscipline dealt with. A straightforward compilation of literature sources or cataloged data is by no means a review paper – especially not in our times when compilations can so easily be carried out. In particular, when composing overview Tables of astrophysical parameters[7] from different sources, one should always make sure that the resulting compilation lists data that are *commensurable* and *homogeneous*, so that their use in statistical studies is permitted. A&A does not publish review papers.

5.5 Errata and Comments

Errata refer to errors in previously published papers, and are normally signed by all coauthors of the original paper. *Comments* are usually not published by A&A, except in very special cases. Three conditions are necessary for a *Comment* to be considered for publication: *i*) it refers to a paper published by A&A, *ii*) it does unambiguously solve the problem or question it raises, and *iii*) its publication will be useful to the community. Some journals publish comments as *Letter to the Editor*.

---

[7]For example, photometric indices, effective temperatures, abundances, …

5.6 Information bulletins and telegrams

A bulletin is a brief news item intended for immediate publication. An example of a bulletin journal in astronomy is the *Information Bulletin on Variable Stars*[8] (IBVS), published by the Konkoly Observatory in Hungary. IBVS publishes short but significant news on variable-star astronomy.

The *Central Bureau for Astronomical Telegrams*[9] (CBAT) issues a telegram journal that is responsible for the dissemination of information on transient astronomical events via the International Astronomical Union Circulars (IAUCs).

The basic difference between these two communication media is the speed (IAUCs being faster) and the publication cost (IBVS not imposing page charges). Telegrams get some editorial attention, but are not necessarily refereed: the degree of refereeing in such short papers depends on the editorial policies of the journal.

5.7 Data papers

Standard journals require concise reporting, hence few journals can accommodate large volumes of experimental data. Some major journals have (or had) associated supplement series, where results of experiments (theoretical or observational) are published. Most data are nowadays published in machine-readable form, and quite some papers in A&A have the associated data (time series, comprehensive Tables, large illustrations, etc.) deposited at the CDS (Centre de Données astronomiques de Strasbourg), see also the paper by Laurent Cambrésy in Part 1 of these Proceedings. Section 14 of A&A, *Online catalogs and data*, offers online publication of large sets of data.

Very few refereed journals accept "data only" papers, *i.e.*, the data without an accompanying analysis. Still, there is a substantial difference between a *data archive* and a *data paper*: whereas the former merely is a set of tabular material with an accompanying readme file, the latter provides the space to discuss the data calibration, to examine all aspects of the error budget, and even to include some scientific results.

5.8 Papers on the design of instruments

Publications on the conception and design of instruments belong to a somewhat similar category as data papers, have a relatively short life time (see also Sect. 5.1.2 in Paper III), and tend to get buried in the archives of observatory libraries. A&A has a Section *Astronomical instrumentation*, with refereed papers on instrumentation. This Section is published online only.

5.9 PhD thesis book

Most universities require the PhD candidate to submit a PhD manuscript. As the pressure to publish forces the student to publish new findings as fast as possible, PhD manuscripts,

---

[8]http://www.konkoly.hu/IBVS/IBVS.html
[9]http://cfa-www.harvard.edu/iau/cbat.html

as a rule, are compilations of published papers, with an introductory chapter that gives the outline of the thesis, and its most important conclusions. As most journals require the transfer of copyrights from the authors (see Sect. 12.2, and Papers II and III), the reformatting of papers in a thesis manuscript could, in principle, be seen as a breach of the copyright agreement. But most journals do not object to reproducing published papers in the PhD manuscripts of their authors.

5.10   Essays

The definition of this form of paper is rather vague, and generally refers to a short interpretive work usually dealing with its subject from a personal point of view of the author.

5.11   Invited talk at a conference

In principle on invitation, such papers give a review of one of the themes of the meeting. Good meetings frequently have several such landmark papers, which are quite often very informative, and certainly are recommended literature for every beginner in the field. My experience is that the term *invited* should not be taken too literally: organizers of meetings often receive an explicit request from a potential participant demanding to be "invited" for a talk.

5.12   Contributed paper at a conference

Such papers, as a rule, are short, and quite often bring a preview of new results that are not yet published. In principle, every conference paper should be able to pass mild refereeing, but in practice, such papers are of widely variable quality, and some would never pass a strict refereeing test. There is now an increasing tendency to include so-called *refereed conference papers* in Proceedings of meetings. The refereeing is then done by the Scientific Organizing Committee of the conference, or by other participants. Needless to say, this approach is a far cry from genuine refereeing, since only the really very bad papers are rejected. The Editor(s), moreover, are in a difficult position to refuse such papers, simply because their content (oral or poster presentation) has been revealed, and the authors risk losing credit if an attendee rushes to print with similar results.

5.13   Conference posters and ticket papers

Poster sessions were created to allow participants to present their results, even if the meeting format only allows invited talks and reviews. The advantage of posters is that participants can view and discuss the material at their own pace, in an atmosphere that encourages direct discussion. But I have seen cases of reproduction or duplication of talks and posters by the same authors at the same meeting – even people postering decade-old stuff. Such duplicates have no value in a curriculum, and such duplication should be avoided. After all, why duplicate a paper if you can come up with other interesting aspects of your research as a *variation* on previously published results? This reminds me of one of my PhD students who proudly told me about his Master-thesis supervisor who was so "clever" that he succeeded in getting the same paper published three times. Frankly, this is more a sign of skill in deception than of cleverness. Section 7.12 in Paper III discusses the problem of paper duplication.

The term *ticket paper* is slang for a conference paper that only serves to get your travel covered. University administrations often require participants to give a talk when they apply for travel support. As such, a short paper or poster is often presented for the sole purpose of acquiring the travel subsidy.

5.14 Book reviews

Books are not really reviewed by a referee, although publishers do ask an external expert to advise them on whether or not to publish a submitted book manuscript. But once a book is published, the publishing company seeks to have the book reviewed by several independent readers, and such reviews are then published in scientific journals. Copies of new books are quite often sent to Editors of journals, who in turn look for an appropriate reviewer. A&A does not publish book reviews.

The principal issues to present in such reviews are content-related criticisms, technical aspects of presentation (such as quality of graphics, occurrence of typographical errors), but also comments on the price and on whether value for money is delivered. There seems to be a clear-cut difference between one- and two-author books (textbooks, monographs) and multi-authored books (mainly proceedings): the latter quite often receive unfavorable judgment because of untidy editing, or as Duerbeck (2005) puts it

> ... VERY SLOPPILY EDITED − OR IS EDITING NOWADAYS NOTHING MORE THAN THE UN-CONTROLLED PROCESSING OF LATEX FILES?

5.15 Publicity papers

Some papers do communicate results, though only for the purpose of bringing to public attention some achievements of a large team or of a scientific consortium. Such papers are typically published in annual reports of institutes, or in observatory magazines.

5.16 Salami papers

The term refers to the not so uncommon habit of reporting results in slices, *i.e.*, instead of presenting all results from a single study in one cohesive paper, the work is partitioned in multiple papers, and submitted to different journals. This technique, for sure, increases visibility in a bibliographic database, but the dilution takes its toll at the moment when a peer-review body asks your *"three most important papers of the last three years"*. More on paper slicing in my contribution on ethics in Paper III.

5.17 Hoaxes

A scientific *hoax* is a fabrication. If the intention of the author is to dupe the reader, we speak of a *fraud* (see also Sect. 7.9 in Paper III). There is a very special type of hoax or *parody* that leads to exposure of corrupt research practices or customs. A most interesting example is Alan Sokal's *Transgressing the Boundaries: Toward a Transformative Hermeneutics of Quantum Gravity*, Sokal (1996). There is a more recent case, where an *Open Access* journal agreed to publish a nonsensical article written by a computer program, claiming that the manuscript was peer reviewed and even requesting that the authors pay 800 US$ in page charges.[10] The debacle led to the resignation of the Editor-in-chief of the journal.[11]

Believe it or not, fabrication of false coauthor names also occurs. One famous example is the most interesting case of immunologist Polly Matzinger, who refused to use passive voice in her papers, and considered that if she is forced to refer to herself as "we", she consequently needs a coauthor. Hence she added the name of her Afghan hound as *Galadriel Mirkwood*[12] in Matzinger & Mirkwood (1978).

Although hoax papers are made to expose misconduct, they are not appreciated, since the form and method of the protest basically violates the principle of mutual trust, as laid out in the last paragraph of the Preamble (page 2). This also applies to inventing coauthors: such fabrications may harm the fabricator in the first place, as Polly Matzinger may have found out.

5.18   Duplicate papers

Paper duplication occurs, not seldomly for commercial or promotional reasons. Sets of papers on one single project (for example, space experiments: see Fig. 22 in Paper III) are often bundled in a special issue of a journal. For celebrating the 40th anniversary of A&A, the Editors published a special anniversary issue as a compilation of 40 papers, which they esteem to have a strong impact in the community at large. This collection appeared as Volume 500 (2009), and contains facsimile duplicates of the original papers, and every such paper was accompanied by a 1- or 2-page commentary by a reviewer. The good thing about this compilation is that an ADS search on any author's name will return only the original paper's bibcode, so that citing these papers will not generate any bibliometric bias. A confusion that may arise when such precautions are not taken is discussed in Paper II (see also Sect. 13.16).

Note also that book Publishers even rename books to fit their markets: for example, Paul Davies' book *The Goldilocks Enigma: Why is the Universe Just Right for Life?* is sold in the US as *Cosmic Jackpot: Why Our Universe Is Just Right for Life*.

5.19   Gray literature

Gray literature refers to instrument and software manuals, applications for observing or computing time, grant applications and grant reports (progress and final reports): in principle all such manuscripts and papers should follow the same guidelines of rigor and

---

[10] http://www.the-scientist.com/blog/display/55756/
[11] http://www.outsidethebeltway.com/archives/science_editor_quits_after_hoax/
[12] Galadriel is a fictional character, and Mirkwood is the name of a fictional forest, in the legends created by J.R.R. Tolkien (1892–1973).

clarity as any of the above-mentioned classes of earnest papers. Dishonest attitudes in grant applications and reports are often severely punished: as a jury member for grant allocations I have witnessed prompt disqualification of applicants who had moved forward the position of their own name in the list of their papers (see also Sect. 11.9). Even upgrading the class of paper (for example selling a poster abstract as a refereed paper) may lead to downgrading an application: some jury members do check!

Press Releases also belong to the category of gray literature, since such texts have, strictly spoken, no author, but a composer.[13]

6 About language

A most important element of the narrative is the language wherein the scientific communication occurs. **Language is the link from the mind of the writer to the mind of the reader** – either instantaneously, or over time. If the world had only one common language, we would have an information-sharing network with formidable power. But there exist approximately 7000 languages worldwide (and several thousand are expected to disappear in the next century), so which language to select for scientific communication?

> LANGUAGE IS NOT JUST ANY CULTURAL INVENTION, BUT THE PRODUCT OF A SPECIAL HUMAN INSTINCT,

says Pinker (1994): humans possess this very special common basic biological skill to learn the processing of information through language.

As Pinker conveys, *"There are Stone Age societies, but there is no such thing as a Stone Age language ... "*. So, basically, almost any language could be used for the communication of science,[14] since all languages consist of a finite list of elements (words or characters) together with a program to construct blocks of information: the syntax for building sentences. But very considerable culture-specific differences exist in communication through language: for example, some cultures express thoughts in an indirect, verbose, affective and tentative way, others communicate in a more direct way. Cultural-specific aspects are evident when someone uses his or her native language, but these aspects are even more detectable when a person expresses thoughts in a language different from the native language.

Today, English happens to be the *lingua franca* in science, and Joli Adams documents this fact in her paper on language editing in Part 1 of these Proceedings. Already in its first year of existence, the Editors of A&A urged authors to use English for communicating their results (see Fig. 1). They point out two prime advantages of using one common language:

1. the work becomes more widely known, and

2. the pool of **active** and **competent** referees becomes larger.

---

[13] See Paper II (Section 4.4) for an example of a paper that has a composer, but no author.

[14] Provided it possesses an extensive technical vocabulary, accompanied by a proper dictionary.

> 1. We wish to advise our French-speaking and German-speaking authors to explore possibilities to write in English in order that the results of their work become more widely known. We now have Editorial help to revise the language of their papers in English if they need it. May we remind the authors that is it becoming increasingly difficult to find active and competent referees capable of reading French or German.

Fig. 1. "First important point" in the very first A&A Editorial (Pottasch & Steinberg 1969).

The use of one common language foreign to so many countries and cultures unavoidably leads to many variants of English, for example Chinese English, Dutch English, French English, American English, etc.: variants[15] where the correct English words are used[16], but that are dominated by the native-language syntax that also permeates the English at the reader's end. And there are also variants of style: scientific style is direct, plain, assertive, context-independent, and brief – in contrast to prosaic or almost any other style that we use on a daily basis. Modern scientific style, however, mostly uses the passive and impersonal voice, and shuns active voice verbs and first-person pronouns.[17]

There are two basic types or classes of languages: *non-tonal languages* (stress languages, like English) that use one dominant *syllable*[18] in the word, and *tonal languages*, which use tone to distinguish words (like Mandarin and Vietnamese).

A tricky situation arises with transliterations from one class to another, like the *Hanyu Pinyin* scheme, which is the most common Standard Mandarin romanization scheme (*Hanyu* refers to the Chinese language, *pin* stands for connection and *yin* for sound): foreign words are phonetically split in syllables that have a meaning in Chinese. Figure 2 shows, as an example, two Pinyin transliterations for the name of the city of Blankenberge, kindly (and independently) provided by two participant students who had never heard about that place before going there. The word is read as *Bù lán kěn bei ge*, and the syllables have meanings of cloth, blue (or orchid), be willing to (or consent[19]), uncle, and grid (or meter). As one can see, the fourth syllable is transposed differently. I have also encountered a kind of "inverse pinyin" in the western transliteration of the first name of Taiwanese student Hsiang-Wen Hsu, who uses the phonetically similar "Sean" as his *callname*.[20] Callnames should never be used in citations or as part of an author name, so this person's bibliometric records in ADS list Hsu, H. as well as Hsu, Hsiang-Wen.

But it really gets catchy when dealing with Asian names of persons. For example, entering the family name Zhang in an ADS search yields almost 14 000 entries, whereas Zhang, L. leads to a list of about 600. The reason is that this name not only is very common in China, but that the Pinyin transliteration for Zhang originates from at least

---

[15]With *variants* I do not mean the national dialects of English.

[16]Though not without confusion: a billion, for example, means 10^9 to some people, and 10^{12} to others.

[17]Notable exceptions are to be found in the weekly journal *Nature*.

[18]A *syllable* is a natural unit in which a word can be analyzed.

[19]The third *grapheme* (a fundamental unit in a written language), and its meaning, are also used for the second syllable in my own surname – referring to a consenting person, perhaps?

[20]A *callname* is a name given in place of the official name, for example Bob for Robert, Liz for Elizabeth, Bill for William, Dick for Richard, and Ted for Edward. A callname is not necessarily a *nickname*.

布兰肯贝格

布兰肯伯格

Fig. 2. Pinyin transliterations for *Blankenberge* (courtesy Wu Yue and Zhang Li Yun).

four different root names. And that is not the only difficulty: Chinese names, traditionally, are formatted differently from Western names. Whereas we write our names in the order given name(s) (first name and middle names) followed by the last name (*i.e.*, family name, or surname), the traditional format in Asian cultures is the other way around, with the family name first (mostly a one-character name – Wu and Zhang in the caption of Fig. 2), followed by a one or two-character given name (sometimes linked by a dash). Nowadays there develops a strong tendency to switch to western format in publications, and even in daily life. This situation greatly adds confusion in reference lists and citations (see also Sect. 9.12), as Jia Wei[21] points out in Qiu (2008):

I REALLY HAVE A HARD TIME SORTING OUT WHO HAS PUBLISHED WHAT.

There are many ways to confuse western names, though. For example, surnames in Spanish-speaking countries come from both father and mother: the first part of the surname is the father's family name, the second surname is the family name from the mother's side (and often only the first letter is mentioned, like in Violeta Parra S.[22]). This format should not be confused with the case of people who have truly composite surnames, nor with cases where people change name by marriage (see also Sect. 9.2). Names transliterated from Cyrillic (Russian and a number of other languages of eastern Europe and Asia) also pose a problem for correct identification, since they do not all use the same Cyrillic spelling: for example, all combinations of Karamazov, Karamazoff, Dostoevsky, Dostojewski, Dostoyevsky, . . . all appear in book translations.

7 The reading process

When reading a text, you are assisted by a kind of mental *daemon*[23] that – in computer jargon – is called a *parser*, a process that **unconsciously captures the hierarchy of the input text and transforms it into a form suitable for further processing**. In linguistics, the parser is a mental program that analyzes sentence structure during reading and, normally, checks for syntax errors at the same time (Pinker 1994).

Let us look at an example from one of my students' manuscripts: *"through linear least-squares fitting an equation that best fitted them was extracted . . . "*. When I read that sentence for the first time, I was mentally slowed down or even stopped after "equation", and again after "them" (*i.e.*, the data). The first part of that sentence contains

---

[21] Associate Dean at the Pharmacy School of Shanghai Jiao Tong University.

[22] Her complete name is Violeta del Carmen Parra Sandoval.

[23] A *daemon* is a computer program that runs in the background.

tautologies: "linear least-squares fitting an equation" and "that best fitted" are useless rep-etitions, since "linear" already specifies the equation, and "least-squares" always implies a best fit. Moreover, "them" is confusing, and can only be understood by remembering the previous sentence. So, a much shorter way of saying the same would be to refer to a linear least-squares fit of two specified variables.

A similar slowing down of the reading pace occurs when previously undefined acro-nyms are used: for example in the same text the acronyms QPO (quasi-periodic oscilla-tion) and AD were used, and the reviewer commented *"It took me a while to associate AD (Anno Domini) with accretion disk..."*. Note that the parser keeps track of bits and pieces of previous sentences (especially acronyms), and that a memory stack of only 6 to 8 posi-tions can be properly maintained. Fortunately, an ambiguity like QPO can be minimized through the context in which it appears – though the writer of a scientific text should al-ways avoid to communicate in context-sensitive statements that imply guesswork by the reader. Note that the parser – just like any other semi-automatic skill – develops with time and, therefore, **intense reading is a prime requirement for adequate development of your writing skills**.

When I read a research paper, for example as a referee, I first read it cover to cover in one fling, and make some brief annotations if needed. This pass is very useful to cool the parser. In a second approach, I concentrate more on the Abstract, the graphs, the Table and Figure captions, and the Conclusions.

One very specialized form of reading is *proofreading* (see also Sect. 10.1), a most important step in the publication process. Proofreading occurs at two phases: several times before submitting a manuscript to the Editor, and then one last time after receiving the proofs from the Publisher. Proofreading occurs already at the earliest stages of writing: in fact, every time you read a few pages of your self-composed text (or text produced by a coauthor), you engage in proofreading, although the proper term for this activity phase rather is *revizing* and *editing*. You can easily put the parser to sleep by too many passes, especially of your own compositions. That is why proofreading is best done by another pair of eyes as well. It also happens that the first Sections are subject to more proofreading passes than the last ones, and that should be avoided, since the last Sections are not the least important!

Proofreading is so important that SCHOLASTIC, the company that publishes the Harry Potter books, has a full-time *continuity editor* who doublechecks the complete proofread-ing. She explains:[24]

> FOR INSTANCE, WITH BERTIE BOTT'S EVERY FLAVOR BEANS, I MAKE SURE IT'S AL-
> WAYS B-O-T-T-APOSTROPHE-S. EVERY FLAVOR IS NOT HYPHENATED, AND FLAVOR
> DOES NOT HAVE A U.

In other words, tight standards of English, as well as of the Harry Potter constructed magic language, are rigidly observed throughout all Volumes of that series.

---

[24]Quoted by Cheryl Klein, Time Magazine, 9 July 2007.

8 Writing a research paper

Writing a paper is a *process*, a chain of gradual steps that lead to an acceptable and pleasing end product. As such, learning to write is a continuous process that demands a lot of time for acquiring the necessary writing skills. The best procedure is to start drafting a paper while the research work is still in progress, recording first the preliminary structure, and then completing those parts that are rather easy to handle: description of the methods, observations, references, Figure and Table captions, etc. One advantage of starting early is that you become familiar with the journal's house style, and almost immediately can see some of the fruits of your labor in typeset format, and that you avoid *writer's block*, *i.e.*, staring at blank pages or at sloppy notes (see Sect. 13.19). Another, more important, bonus is that starting early may also lead you to follow possible sidelines, even to a spin-off paper derived from your main work. Such a by-product paper is an excellent item to present at a meeting as a poster, to show new findings in the context of the core work, without needless duplication of material that has already been published elsewhere. In such a way, you can **make publicity for your core work, without violating copyright rules, nor creating copy-and-paste duplicates**.

There are four indispensable basic requirements for producing a good paper:

1. write clearly,

2. write accurately,

3. be brief (avoid verbose and pompous styles), and

4. build a logical structure: the train of thought should be logical, and avoid a winding and repetitive course in the suite of ideas.

Besides these basic requirements, there are a number of issues to absolutely avoid, see Section 11.

A very important point is to learn mastering *the exact word in the exact place*. Table 1 comprises sets of nouns, verbs and adjectives that look like synonyms, but do not always have exactly the same meaning in context, by which these words are often incorrectly used. It is very good practice (also for native English-speaking authors) to consult a *thesaurus* – a list of words grouped together according to similarity of meaning – rather than always fall back to translation by means of a dictionary.

A guaranteed way to use the wrong word in the wrong place is grandiloquence, *i.e.*, the excessive use of verbal ornamentation and pompous language. Or giving impressive numbers with incomplete information content, like the digital-camera manufacturers' marketing hype of playing with zoom factors and resolution specifications. I recently reviewed a paper where the authors stated that they used a 390 150-pixels CCD camera, whereas the more modest-looking 715×510 gives the reader much more information on the chip's dimensions and size.

Some verbs like *look, believe, seem* and *think* should be used with great care: avoid writing that "X thinks ...", because you cannot really know what X thinks or believes. One of my postdocs liked to state "the data look good ...": *good* is not a generic synonym for *excellent*, and covers *pleasing* as well as *dependable*, hence a more quantitative and

Table 1. Groups of words that look like synonyms, but are often incorrectly used.

error, uncertainty, mistake, defect, flaw, blunder	scatter, noise
dimension, size, extent, extension	trend, pattern
regression, correlation, fit	chance, probability
confidence, significance	amplitude, range, power
standard star, comparison star	flux, intensity, luminosity
seeing, scintillation, scattering	precision, accuracy
color index, filter, passband, color	robust, stable
standard, classic, normal, default	parameter, variable, estimate
misbehavior, misconduct, misdemeanor	symmetry, isotropy
define, postulate, speculate	bias, residual
parameter, observable, factor	minimum, lower bound
colloquium, symposium, conference	maximum, upper bound
presentation, talk, lecture, seminar, class	typeface, font, type
supervisor, mentor, coach, patron, boss, sponsor	retention, secretiveness
do, perform, conduct, carry out	great, large, big
period, frequency, mode, harmonic	simple, elegant, elementary
fact, observation, measurement	randomness, entropy, chaos
model, doctrine, theory, hypothesis, mechanism	duplication, redundancy
supposition, proposition, assumption, premiss	mean, average
example, metaphor, conjecture	consideration, thought
prove, claim, maintain, demonstrate, verify, contend	invention, discovery, insight
assert, avow, support, advise, suggest, establish	significant, relevant
predict, anticipate, forecast, foresee	whole, complete, entire
instructions, guidelines, requirements, rules	mention, cite, quote, refer
calculate, compute, reckon, count	emulate, simulate
authenticity, integrity, honesty	code, algorithm, program
expect, believe, suppose, estimate, guess, think	intercept, zero point
excellent, good, satisfactory, acceptable, sound	pleasing, dependable
tutorial, guide, manual, treatise, memoir	manuscript, paper, text
absolute, relative, differential	ask, suggest, recommend
explain, understand, apprehend, comprehend	inconceivable, impossible
confirm, affirm, corroborate, validate	induce, conclude
entail, implicate, impose, imply	luck, chance, serendipity
enormous, immense, indefinite, infinite, innumerable	refute, overthrow
copyright infringement, (self-)plagiarism, paraphrase	repeat, replicate
jargon, terminology, nomenclature	construe, interpret
capture, acquire, sample	analyze, reduce, process
discern, distinguish, recognize, scrutinize	utilize, use, apply
invent, contrive, formulate, imagine, devise	revise, review, referee

less emotional expression is to be preferred instead. Qualifying terms, such as "likely, probably, perhaps" are also not very recommended ways to make your point, unless the point is so vague that "perhaps" it is not a point at all. Modifiers like "possibly, very" weaken the point even more, as well as "it is clear" or "it is inconceivable", etc. Declarations that are made empathically, like "there can be no doubt", should also be avoided, as they suggest that no supporting evidence is necessary.

Look out when using words with a very strict meaning in mathematics and physics: sometimes, a quantity or a well-defined parameter is used in a context so vague, that the reader has no clue of what that quantity or concept really represents. One common error, for example, is to call a frequency a *mode*, before even knowing that the frequency is associated with stellar pulsation. Another one is the symbol σ, which is normally used for the standard deviation[25] of a single observation, associated with $\pm$ pointing to the standard error of the mean, identifying the range of values that encloses the true value of the measured quantity. But in past times, $\pm$ was often used to indicate probable error (p.e.). Hence, avoid writing σ, or "the sigma value", but specify what σ means. And do not assume lightly that an error distribution is *normal*[26] (or Gaussian), which strictly means that the expectation of every variable is the same as the expectation of any other in the sample (and each is equal to the mean or average), and implies uncorrelated values. Another frequently occurring confusion is *variance* – *i.e.*, the square of the standard deviation – where a biased or an unbiased estimate can be implicated. And in counting samples, *ten or more* is not exactly the same as *more than ten*.

An example of an ambiguous word is *bug*, meaning either an insect, or a hidden microphone, or an error in a computer program. Another example strictly belongs to computing: the symbol NaN[27] is mostly produced as the result of an invalid operation, and this symbol should not be used for missing values in Tables, unless it is specified that the symbol stands for a missing number and not for the result of an unsound computing operation (see also Paper II).

Non-native English-speaking authors[28] should always look up the meaning of a word whenever in doubt, especially when words with different meanings have almost the same pronunciation (so-called *homophones*, for example "whole" and "hole"). Words that are considered synonyms in your mother tongue, may have a different connotation in English (like large/great/big, intriguing/interesting, claim/maintain/prove, or listen/hear, see the footnote on page 5).

Beware also those words that are called *false friends*: pairs of words in two languages that look or sound similar, but differ in meaning. A very frequent misunderstanding involves *eventually* (meaning an unspecified later time), which matches the French *éventuellement* (possibly). Other examples are *data* and *date*, *curriculum* (syllabus) *versus* CV[29], and the mixup of *to use* and the French *user* instead of *utiliser*.

---

[25] Standard deviation is a measure of the dispersion of a data set.
[26] *Normal* is a misleading word, and has a different meaning in "normal distribution" and "normal equations".
[27] NaN: Not a Number.
[28] Sometimes referred to as *non-native users of English* (Salager-Meyer 2009).
[29] *Curriculum vitae*.

Beware also of constructing English words by simple logic: someone who theorizes, can be called a theorist, but someone who socializes is not necessarily a socialist. And one who organizes is an organizer, not an organist (though some organists do organize organ concerts). Do not forget that language is a continually evolving communication system with strict rules – though these rules do not lead to 100% logic consequences.

Be very careful with translation, and never write your paper first in your native language for translation afterwards (by yourself, or by a friend): **there is a real danger that the translation process changes your message**. As an anecdotic example, I refer to an instruction manual of an electric door bell, which stated that installation was easy, and that only "three sons" were needed: the automaton-like translator had mixed up the French word for cables (*fils*) and for sons (*fils*) without attention for the coherence of the message. This kind of machine translation is offered by internet search engines, and can lead to very strange sentences when applied to scientific texts, as the following example illustrates.

Li *et al.* (2007) published a paper entitled *A new study of the long-term variability of the SX Phoenicis star CY Aquarii*. The Google translation of the abstract goes: *"By phoenix SX-type variable star CY Aqr was determined by observing two new light changed a great time, ... Combining the published data points fitting reveals a new formula for calculating a great time"*. Obviously, the constellation was identified with a mythical being, and time of maximum was rendered as "a great time".

In order to avoid translational gaffes, when quoting a text in translation, it is advisable to give the original statement together with your translation (as is done in the footnotes on pages 4 and 20). Non-native English-speaking authors should also consider the advantages of using English-language versions of computer operating-systems, of viewing non-dubbed English movies, and of reading English books in their original editions.

9   The structure and organization of a scientific paper

People often mix up *manuscript* and *paper*. A manuscript[30] is a document submitted to an Editor (or to be submitted in a not too distant future). A paper is a manuscript accepted for publication – that is, as soon as it has been stamped "accepted", or has received a Digital Object Identifier[31] (DOI®) from a Publisher.

We have seen that words are the raw materials, and that language syntax gives the sentences their meaning. In a paper, sentences are grouped in paragraphs, and paragraphs become the elements of Sections. Most scientific papers have a very similar structure in a well-tried format suitable to efficiently transfer facts and interpretations of facts. Papers are mostly organized in Sections according to the so-called IMRaD model, where the acronym stands for **I**ntroduction, **M**ethods (observations, computation, theory), **R**esults and **D**iscussion (and Conclusions).[32] The A&A website states about this common model:

---

[30]Etymologically, the term refers to a hand-written document, and a non-manually reproduced text was called a typescript.

[31]The Digital Object Identifier System identifies objects in the digital environment (*i.e.*, publications) with an internationally standardized unique DOI name, see http://www.doi.org/. The footer of the title page of each paper in these Proceedings holds the DOI.

[32]This scheme is not applicable to gray literature.

THIS IS A WELL-TRIED FORMAT; AUTHORS SHOULD HAVE GOOD REASONS FOR DEVIATING FROM IT. THE GOAL OF A SCIENTIFIC PAPER IS NOT TO IMPRESS THE READERS BY POETIC LANGUAGE BUT TO TRANSFER FACTS AND NEW INSIGHTS AS LUCIDLY AS POSSIBLE.

An example of non-IMRaD is the paper by Oort (1970) *The Density of the Universe*: this paper has an abstract, but no further structure. This is only admissible for very short, and very important papers.

Note that the IMRaD model does not dictate that each element of the acronym must necessarily be composed after the preceding one. My experience is that the **M** and **R** Sections are the easiest to start with, whereas **I** and **D** are best finalized at a later stage. In the following Subsections we deal with each structural element of the manuscript.

9.1   Title, subtitle and running title

The title should be specific, and as brief as possible, at the same time it should be attractive. If a title is long, then the author should provide an abbreviated *running title* for the page headers. Avoid neutral titles (beginning with "A study of ..."), refrain from excess words, do not use abbreviations (unless they are very well known, for example, LMC for the Large Magellanic Cloud), and prevent grammatical errors and typos in the title. Finalize the title when you finish the paper. A subtitle is used when the title is too long to be clear, or to indicate that the paper is part of a sequence of papers.

I once saw a booklet entitled *On BH's and GW's*. It took me a moment to realize that the compilation was about black holes and gravity waves, especially because the plural of BH and GW should have been written as BHs and GWs – without apostrophe[33]. Such a title, though brief, is not clear, as the reader needs reflection to understand it. The champion of brevity and wittiness, sure enough, is "*M* 32 ± 1" by Lauer *et al.* (1998).

Question marks in the title can also lead to confusion, for example *Stephan's Quartet?* (Allen 1970), which implicitly refers to Stephan's Quintet. Or *Mode Identification through Photometry: Full Colours?*, where the appellation "Full Colours" refers to recognition of levels of achievement or service,[34] see Sterken (2000) – an intended pun to lead a paper that critically discusses some deficiencies of mode-typing photometric methods. A similar strategic ambiguity is to be found in the title of Watson's (2007) book *Avoid Boring People*, where boring can be read in two ways: as the adjective meaning non-exciting, or as the gerund of the verb "to bore". Such confusion is mostly undesigned, and may pose a real problem for people acquainted with languages where the usage of a verb as a noun[35] is uncommon.

A famous counterexample of a very clear and most complete title is by Lemaître (1927), who formulates his hypothesis in a most explicit way already in the title of his work[36]: *A homogeneous universe of constant mass and increasing radius, validating the*

---

[33] An apostrophe expresses a relationship through the genitive.

[34] in athletics, schools, clubs, etc.

[35] A quite unique aspect of the English language.

[36] *Un Univers homogène de masse constante et de rayon croissant rendant compte de la vitesse radiale des nébuleuses extragalactiques.*

radial velocities of extragalactic nebulae (see also Paper III). However, beware to never promise more in the title than it delivers!

9.2   Authors

Who should be coauthor, and in which order author names must appear, are issues many young (but also some not so young) writers struggle with. Virtually nobody carries out research without any help at all: there are several classes of people who provide assistance: scientific contributors, technical contributors, administrative and financial contributors, etc., and it is evident that some of these contributions, **if substantial**, deserve coauthorship. There are no absolute rules, except for the general expectation that the order of names should be morally upright and correct, *i.e.*, **that there is a correlation with delivered effort or labor**.

Though it may sound far-fetched, typos in author names – even the same surname appearing twice in the same author list – do occur! Typos most frequently appear in names with character accents (*e.g.* č, ç, œ...) and in composite names. The many ways of using the prefixes *van*, *le*, *du*, ... can lead to problems with search engines. Take as an example van Genderen, which could bring about Van Genderen, Van genderen, and Vangenderen. Any of the first three orthographic forms leads ADS to more than 250 abstracts coauthored by Arnout van Genderen, whereas "Vangenderen" generates no hits at all. A similar, but reverse, situation occurs with Du Bois, du Bois, du bois and "Dubois": only the last format leads to the abstracts of the papers of Patrice Dubois. And surnames are not everywhere processed in the same way when it comes to alphabetic order: names with prefixes are in some countries ordered according to prefix, and in other countries according to the principal component, for example "Genderen, van" and "Zach, von", see the name index list at the end of this book.

Beware also that many people have several first names, and that some of these are composite: there is a difference between Jean Michel (J. M.) and Jean-Michel (J.-M.). To illustrate a real-life case: I have coauthored four papers with Chinese author Liu Zongli, and the ADS search supplies four different identifications: Liu Zongli, Liu Z., Liu Zong-li and Zongli, L. And to make the confusion complete, my university's library database lists several works under "Sterken C.", and one book by "Sterken Chr." – simply because the Publisher decided, without any consultation, to adopt Chr. for the initial of my given name[37] in Sterken & Manfroid (1992).

There is an additional complication with gender: in some societies, marriage coerces females to adopt the name of their husband. So, Hillary Clinton was Hillary Diane Rodham before her marriage – Rodham was her maiden name – and she is sometimes designated as "Hillary Clinton *née* Rodham". But she is often referred to as Hillary Rodham Clinton, though some would call her Hillary Rodham-Clinton and others will refer to Hillary Clinton-Rodham, a really awful situation if she were a scientific author, and perhaps to be only surpassed by Irène Joliot-Curie[38]. Not to speak of what happens when a

---

[37]What prompted several colleagues to address me as Christopher.

[38]Irène Joliot-Curie was the daughter of Marie Skłodowska-Curie and Pierre Curie. In 1926 she married the physicist Jean Frédéric Joliot, and they both changed their surnames to Joliot-Curie.

female scientist divorces or marries again, implying a complete reset of the citation count. And how to handle same-sex marriages?

The only advice I can give to female scientists is to simply retain their birth name (and initials), and take an example at the American suffragist Lucy Stone (1818–1893), who was the first to claim the right not to use husband family names. To illustrate the depth of the problem, I refer to Merton (1995) on the so-called *Thomas Theorem*,[39] and quote from Dorothy S. Thomas' letter of September 10, 1973 addressed to R. K. Merton:

> I ASSURE YOU THAT I WAS A BORN THOMAS AND THEN MARRIED A THOMAS WHO WAS NO RELATION. I WAS AN ARDENT LUCY STONER[40] AND ALSO SWORE I WOULD NEVER CHANGE MY MAIDEN NAME WHICH I DIDN'T.

The proper way for these authors to sign would be Thomas, W. I. & Thomas, D. S., and the correct citation would be Thomas & Thomas (1928)[41].

An example from astronomy is Maunder & Maunder (1908): *The heavens and their story*, by Annie Scott Dill Russell and Edward Walter Maunder. Annie Russell (1868–1947) was a solar astronomer, and one of the very few 19th-century professionally-employed female astronomers in Britain (Brück 1994). In 1895 she married Edward Walter Maunder, and their book was coauthored "Annie S. D. Maunder & E. Walter Maunder" – though the front page gives "A. & W. Maunder". ADS always lists his initials as E. W. (both initials referring to his given names), whereas she is either referred to as[42] "Maunder, W., Mrs." or "Maunder, A. S. D.", where the last initial does not refer to a given name, but to her mother's surname Dill. This brings up another aspect of quoting given names: though born Edward Walter Maunder, he never used the first given name, and always signed as E. Walter Maunder. Whatever the reason for dropping the first name, if it is done when signing a paper, **it should always be done in the same way** – unless you do not mind bibliometric mix-up.

This brings me to abbreviations of given names, for example Thos. for Thomas and Wm. for William – an old tradition that was introduced for saving a few characters in print, but that causes more confusion than savings. And, as said already on page 12, common callnames, nicknames and diminutives all contribute to confusion in reference lists, and should thus not be used.

Finally, a last word on gender and given names: it is often far from obvious to guess the gender of a person on the basis of one or more first names, especially when dealing with names in other cultures. Some boys' names look deceptively similar to girls' names in another language (for example Cecil and Cécile), whereas Chris, Scott, Kim and Andrea are unisex names that serve males as well as females.[43]

---

[39] Sociologists William I. Thomas and his wife, Dorothy S. Thomas, coauthored on an axiom in the sociology of knowledge: *"If men define situations as real, they are real in their consequences"* – though for a long time all credit solely went his way.

[40] Females who insist on keeping their maiden name are called *Lucy Stoners* ever since.

[41] This reference is not included in the reference list because I had no access to this work.

[42] Maunder (1902, 1904).

[43] As recorded by the US Census Bureau http://www.census.gov/genealogy/names/

9.3 Affiliations

Author affiliation is not just a matter of recording contact information: affiliation information is increasingly used to rate universities and journals, and the outcome of an affiliation index can be decisive for an institute's financing. It is, therefore, very important to use correct and unequivocal affiliations for all coauthors. The following points need special attention:

1. make sure to use the same format for all coauthors;
2. beware of abbreviations and acronyms;
3. be consistent and use only one language inside the address (for example Brussel-Bruxelles-Belgium would involve three languages in one address line);
4. verify if the English translation of your institute's name is recognized by the search engines, and if it is not, then stick to the original;
5. if you mention "on leave", then be specific and correct.

The last point may need some clarification. In my editorial practice, I once encountered a PhD student stating his affiliation with an additional *"On sabbattical leave from . . . "*. A *sabbattical*[44] is a prolonged absence in order to fulfill a very specific goal, *e.g.*, writing a book, and sabbaticals are typically for one year following six years of full-time teaching or administrative activities, and certainly do not belong to the scope of a regular PhD training.

9.4 Abstract

The abstract should provide maximum information with minimum words, and cover the following elements:

1. *WHY* was this research undertaken, and what is the objective of this study;
2. *HOW* did you do the research (observations, theory, calculations), and
3. *WHAT* are the new results, and what do these new results mean,

and should be comprehensible without reading the paper. The abstract is not part of the paper and should not make reference to it. It is preferably written in the third person, and draws attention to all new facts in the paper. Avoid using abbreviations and acronyms,[45] or listing references in the abstract: if referring seems inescapable, then simply give the complete reference, including journal Volume and page numbers. Grammatical errors and typos in the abstract should not occur, and any kind of review matter is out of place there.

Remember that the abstract is the first thing someone reads after reading the title (especially when consulting journal archives, where reading the complete paper may involve access and download permissions), and as such, the abstract is perhaps the most important part of the paper. Some journals do not include all back volumes in their online archives, and comprehensive online collections of journals (like SPRINGERLINK[46] or

---

[44]From Greek *sabbatikos* and Latin *sabbaticus*.
[45]Unless they are very common, like FAQ used in the Abstract of this paper.
[46]www.springerlink.com/

SCIENCEDIRECT[47]) offer publisher-imposed restricted access modes,[48] and require the reader to pay for downloading the full paper. Since the abstract is mostly available, a complete and well-composed abstract will help the reader to decide whether it is worth paying for the download.

To help you optimize your abstracts, A&A encourages the use of structured abstracts (see the editorial published by Bertout & Schneider 2005). Just like a traditional abstract, a structured abstract summarizes the content of the paper, but it does make the structure of the article explicit and visible. For doing so, the structured abstract uses headings that define several short paragraphs. Three paragraphs, entitled *Aims, Methods* and *Results*, are mandatory. When appropriate, the structured abstract may use an introductory paragraph entitled *Context*, and a final paragraph entitled *Conclusions*.

The objectives of the paper are defined in *Aims*, the methods of the investigation are outlined in *Methods*, and the results are summarized in *Results*. The heading *Context* is used when needed to give background information on the research conducted in the paper, and *Conclusions* can be used to explicit the general conclusions that can be drawn from the paper. The Editors of A&A, though, suggest that authors who prefer the traditional non-structured form are invited to implicitly follow the logical structure embedded in the proposed structure. Note that the style of the abstract should be pleasing. Bertout & Schneider (2005) demonstrate how structured abstracts may look by giving an example from an observational paper, and a second one from a theoretical paper in cosmology.

9.5   Key Words

Key words are used for indexing, and A&A allows a maximum of six key words to be listed immediately after the abstract. These words are to be selected from a list that is published each year in the first issue in January,[49] a listing that is common to the major astronomical journals.

9.6   Introduction

The introduction should clearly state why the study was undertaken, and place the research in a broad context by referring to previous works of relevance. The introduction should not contain the conclusions. Some authors tend to expand an introduction into a review paper by itself; this should be avoided, it is better to refer to papers in the well-established review journals. At the end of the introduction the outline of the paper may be described as a road map of your work, but spoiling space for presenting a list of Section headings at the end of the introduction (as I do in Sect. 1) is to be avoided when space is limited.

The introduction is also the place to mention if preliminary results or related conference papers of yours were already published or submitted. Give a clear statement of the problem you study, and present a brief literature review that cites the most relevant papers

---

[47]www.sciencedirect.com/

[48]Either embargo, *e.g.* a 360 day delay of availability of full text articles, or access to only part of the archive.

[49]Keywords are subdivided into broad categories, see www.aanda.org/content/view/136/184/lang,en/

related to your work, but avoid irrelevant citations (especially of the few people you personally know). Only give references to papers that you have really read, or at least have seen, and never cite a reference because the paper or software manual you read also cites that paper! Wild citing changes the citation history of a work, and this helps no one (see Sect. 18 in Paper III). If you refer to someone, refer to a specific and important paper that is of direct relevance to your work, not just to a minor poster or abstract.

In the introduction, but also throughout the rest of the paper, a proper paragraph structure should be maintained. A paragraph is a set of sentences in which one and only one main point or idea is handled. It also includes directly-related sentences with details and information that support the paragraph topic.

Finalize the introduction after finishing the discussion and conclusions, and check if all acronyms have been properly defined.

9.7 Methods, observations, computations and theory

In this Section you describe your methods in such a way that the description gives all elements needed to allow a full understanding and possible experimental or computational replication or verification of your work. Do this in a concise way, list observing logs in a Table (with maximum information in a minimum of space), and if your paper discusses multiple targets, avoid repetition of redundant descriptions for every object discussed.

9.8 Results

Point out how the data look (trends, new effects, frequencies, . . .), but do not give interpretation of the results at this stage: use the data to state only the bare facts, without making inferences. Describe experimental errors, and state the accuracy of the results (but avoid tabular and graphical redundancy, since Tables and Figures illustrating the same results are mostly not accepted by journals). Do not oversell nor underreport what you found, and always question your assumptions. As Richard Feynman (1985) explains:

> . . . IF YOU'RE DOING AN EXPERIMENT, YOU SHOULD REPORT EVERYTHING THAT YOU THINK MIGHT MAKE IT INVALID — NOT ONLY WHAT YOU THINK IS RIGHT ABOUT IT: OTHER CAUSES THAT COULD POSSIBLY EXPLAIN YOUR RESULTS; AND THINGS YOU THOUGHT OF THAT YOU'VE ELIMINATED BY SOME OTHER EXPERIMENT, AND HOW THEY WORKED — TO MAKE SURE THE OTHER FELLOW CAN TELL THEY HAVE BEEN ELIMINATED. . . . IN SUMMARY, THE IDEA IS TO TRY TO GIVE ALL OF THE INFORMATION TO HELP OTHERS TO JUDGE THE VALUE OF YOUR CONTRIBUTION; NOT JUST THE INFORMATION THAT LEADS TO JUDGMENT IN ONE PARTICULAR DIRECTION OR ANOTHER.

9.9 Analysis, discussion and conclusions

Here comes the analysis and interpretation of the results reported in the previous Section. It is very important to keep this Section separate from the foregoing. Compare your results with published work, and point out limitations and uncertainties of your own work (see Feynman's principle in Sect. 3 of Paper III). Always translate the observational and computational accuracy to error budgets *in the physical domain*, and avoid confusing the error of a numerical fit with the error on the result (see Paper II). Give suggestions for improving your results, mention competing explanations and contrary evidence. Your conclusions must absolutely stand out in this Section.

Concerning the reliability of your results, there are two concepts that are being confused all the time:

1. *precision*: how finely a result reproduces (statistical estimate or observable), and

2. *accuracy*: how close a result is to the true value.

Young (1994) puts it like this:

> BY "PRECISION" IS MEANT THE REPEATABILITY OF A MEASUREMENT, USUALLY UNDER FIXED CONDITIONS. ON THE OTHER HAND, "ACCURACY" MEANS THE ABSENCE OF ERROR, AS MEASURED AGAINST SOME EXTERNAL STANDARD, SUCH AS A SET OF STANDARD STARS."

See Figure 18 in Paper II for a graphical representation of the difference between precision and accuracy.

Whereas both concepts can be reasonably well described in statistics, observation and measurement, it is not evident to disentangle them when presenting models that include many input parameters, especially when these rely on calibrations. Take, for example, the light-curve modeling of a binary star that results in almost two dozen output orbital and stellar parameters: a vast number of these quantities are frequently reported with unrealistically high accuracies.

9.10 Acknowledgments

To always give credit and acknowledge the help of others, is a matter of scholarly courtesy. It is also a matter of honesty and fairness towards your supervisor, a colleague, or a referee. The most important point here is that acknowledgments need both be fair and complete. In short, **acknowledgment is an obligation**.

Avoid pages-long acknowledgments thanking everyone who crossed your path. And be sincere and do not exaggerate: if you had a not so pleasant experience with a supervisor or assistant-supervisor, then do not thank them with extreme zeal, but find a polite way to transfer the obligation (unrestrained flatter may even insult your other colleagues). By all means avoid phrasing it like *"The author wishes to acknowledge the work of Dr. X towards formulating the concepts and writing the first draft of the manuscript"*, or *"I would like to acknowledge the help of A, B, and C. They obviously bear no responsibility for the contents of this article"*,[50] which are confusing, or may very well be interpreted as mockery. And, when mentioning people, give their complete names (Jim, John and Jane will appreciate if their surnames are also given).

I have experienced an acknowledgment of the style *"The first author wishes to thank B for useful comments on the manuscript"*, at the same time omitting C who performed a critical reading of the paper long before B came on board. Such negligence can be

---

[50]These quotes were taken from real publications.

interpreted by C as if his/her contribution was of no impact at all, or that this is a case of an "all take and no give" thing. Such situations may result in crushing the collaboration.

As you are not supposed to forget a coauthor in the author line, you should also not forget to include the names of all persons who contributed to the writing of your paper. If a language professional (editor or translator) has helped you with the manuscript (especially with grammar, punctuation, and style), you should acknowledge that assistance. But there is a catch here: acknowledging language assistance implies that you follow the language editor's advice, and that you do not introduce new errors when correcting the manuscript, nor that you add any last-moment sentences that the language corrector has not seen. Acknowledging a language expert, while producing a poor product, is more of an annoying affront than a thanks.

Thank the provider of the equipment that you used (library, telescope, computer facilities, etc.), but never feel obliged to add all directors and associate directors of the place to the list of authors, as is explained in Section 13.1. Explicit mentioning of research and travel grants is mandatory, and some agencies require the use of standard expressions, and also the inclusion of a list of contract numbers.

The situation is not unlike the one in the performing arts, in particular movie productions, where the important contributors – major actors and the director of the project – are listed up front on the movie poster, while lesser contributors – technical assistants, figurants and supporting bit-players are listed in the trailer at the back. Spectators, of course, do not read every name in the trailer, and the same is true for the reader of a scientific paper who will skip the acknowledgment paragraph, but it is nevertheless important for those people who contributed that it is mentioned in which part they have been involved. If you publish a volume of text, you may add a sentence like "cover design by A, artwork by B, typeset by C, printed by D Press, published by E...".

Do not assume that acknowledgments are not read. I often do, and I find that it gives me an additional insight in the sociological aspects of science. For example, I was struck by the statement

> I ALSO WISH TO RECORD MY INDEBTEDNESS TO KEVIN PRENDERGAST FOR DISCUSSIONS (IN 1979) ABOUT ACCRETION DISK PROCESSES THAT FIRST SPARKED MY INTEREST IN THE PROBLEM.

(Shore 2009), that very well illustrates the long-term aspect of this very wide service community that science is, or should be.

9.11 Dedications

A dedication is mostly a tribute to one or more persons, and is rarely used in scientific reports. But it is quite often included in a PhD thesis manuscript (*"in memory of..."* or *"to my parents"*), or in a book. A dedication not necessarily needs to include the ground for the statement, but giving a reason helps the reader to understand the message. For example, Tukey (1977) lauds two colleagues in the following terms:

> DEDICATED TO THE MEMORY OF CHARLIE WINSOR, BIOMETRICIAN, AND EDGAR ANDERSON, BOTANIST, DATA ANALYSTS BOTH, FROM WHOM THE AUTHOR LEARNED MUCH THAT COULD NOT HAVE BEEN LEARNED ELSEWHERE.

9.12 References and citations

Reading and writing are intimately connected. Citation, from the author's point of view, is the practice of referring to the work of yourself and of other scientists. Sweetland (1989) distinguishes four additional functions of citation from the reader's standpoint:

1. *information retrieval*: citing and cited papers are related by subject;

2. *document retrieval*: through (inter)library services and downloads;

3. *scientometry*: the ranking of journals, and

4. *bibliometry*: the use of citation counts to determine a scientist's productivity.

These functions require that citations are accurate, complete and correct.

Each citation in the text must entail a reference at the end of the paper, and *vice versa*,[51] each reference must lead to at least one citation in the text. Cite and quote correctly and accurately, and only give references to papers that you have really used for the paper that you are composing. Sometimes it is not possible to give a formal reference to a published paper – for example when receiving information from a visiting colleague, or through email messages and conversations in electronic discussion groups – then the proper way of citing is the author's name and year, followed by *"personal communication"*. If you use unpublished graphics, always cite the original sources of the Figures, and add *"courtesy of ... "* in the caption.

Two main systems for citing and referencing are in use: the HARVARD SYSTEM, with author names and year in the text, and references in alphabetic order at the end (like in all papers of this book), and the VANCOUVER STYLE, with author-numbers in the text with references in footnotes or numbered at the end. A&A uses the Harvard system. Each book or journal has its own house style for the half a dozen basic fields that comprise:

1. Name(s) of author(s)

2. Date (year)

3. Title of paper

4. Name of journal

5. Volume number

6. Page number (or range).

It is not mandatory to include the title of a cited journal paper,[52] but for books and reference works it is required. Note that with six fields, there are in principle 720 possible combinations, and allowing for some variance in the format of the author names (leading or trailing initials, etc.) one easily arrives at millions of possibilities! Hence, some standardization is imposed, for instance by using internationally recognized abbreviations for journals, by keeping punctuation to a minimum, and by avoiding underlined or embolded elements. As publishing space is tight, cited papers with more than four authors are usually listed in the reference lists as author1, author2, author3, *et al.*

---

[51] Latin phrase meaning *the other way around.*

[52] In this book, all titles of papers are given in the list of references.

The fact that different citing and referencing systems are in use in our field of science, leads to an unanticipated bonus for Editors and referees: I have been able to trace several duplicate papers, simply because the authors reprocessed the same manuscript using a different style or class file, but forgot to adapt the formatting in the reference lists!

Typos in the reference list should not occur, especially in the author names and in the bibcode-related parameters. Three frequently occurring aspects of sloppy typing can really be a nuisance to the reader, in particular for on-line papers: wrong coding or even omission of *diacritics* (see Sect. 13.15), inclusion of control characters in URL addresses,[53] and adding spaces inside, or alien characters (bracket, dot, comma, colon, . . .) immediately before or after an URL.

9.13   Postscript and Appendix

A *postscript*[54] is a note or series of notes appended to an article or a book. It is reserved for a very special remark or statement (see the postscript on page 59). An *appendix* contains supplementary technical matter in tabular form, usually attached to the end of a paper or book, not to confuse with the Postscript described previously (for an example of an Appendix, see the paper by Laurent Cambrésy in Part 1 of these Proceedings).

9.14   Index and glossary

An *index* is a list of words or headings and associated page numbers, and appears at the end of a book (thesis manuscript, textbook, conference proceedings), and is not included in a research paper in a journal. One can have author index, subject index and name index, but also an object index (for example with celestial object identifiers).

A *glossary* is an alphabetical list of technical terms with definitions of those words, and is usually presented as one of the appendices at the end of a book or thesis manuscript.

10   Submitting your manuscript

10.1   The very last step: deep proofreading

Proofreading is not a natural talent or instinct, but an acquired skill, and you must learn to master it by developing an "eagle eye" for your own mistakes. There are, basically, two modes: *proofreading for content* and *proofreading for error*: the former can only be done in the reading equivalent of the listening mode of hearing (see the footnote on page 5), the latter can be done in a more relaxed semi-automatic background-processing mode .

Proofread very slowly, one word at a time, and read what is written on the page, and not what you think or remember is there. The very last double-check in proofreading

---

[53]URLs often contain non-ascii characters, therefore the URL encoder converts the URL into a valid ascii format by replacing the non-ascii characters with % followed by two hexadecimal digits. Such control characters should not appear when citing URLs. ascii stands for *American Standard Code for Information Interchange*.

[54]Postscript is a trademark used for a programming language optimized for printing graphics and text, but *a postscript* – in the present context – refers to *post scriptum* (abbreviated as PS), a sentence or a paragraph added after the main body of a letter or text. It is a kind of afterword or epilog.

should be title and abstract. Always ask someone else to check the title page of your thesis manuscript, and especially check for omission of blank spaces and the wrong use of the hyphen, because these are the most painful mistakes to live with. Anyone experienced in proofreading knows that it is much more difficult to detect errors during on-screen reading than when working with a printout. Basically, there are various errors to be corrected while doing the final proofreading, such as:

1. mixture of American and British style (like analyse, analyze);
2. typos and mistakes of things you know, but do not see;
3. language errors unknown to you, and
4. style and syntax errors.

The first three kinds of errors can be fixed with a spelling checker, but the last variety is more tricky, especially since a mistyped word or verb may be very well correct in a different context.

After reading over your manuscript several times, you will somehow have become blind to errors, because you subconsciously dictate rather than read. Ask a close friend or a colleague who is not involved in your work, to proofread your entire manuscript. It is good practice to look for a *sparring partner*[55], and this is how I operate for some of my manuscripts (see the acknowledgment at the end of Paper III, and in the Preface of this book).

There are several beliefs about when to stop proofreading. One is the fatal *"no matter how many times you read the manuscript, there will always be undetected errors, hence proofreading is useless"*. The other is what I overheard my own supervisor saying to a novice student *"we have done enough now, and any remaining errors will be flagged by the referee"*. Both attitudes abuse the system.

You should always be aware that, when all the experimental and analytical work is done, and when the manuscript has been circulated umpteen times to all coauthors, one must stop at some point, and refrain from going on forever changing what has been changed several times already. This phenomenon is quite typical for a situation with "too many bosses", *i.e.*, the student having to take to heart the advice of too many senior coauthors. It is good to be aware of the so-called *Law of Diminishing Returns*: a general law of economics proposed by Malthus (1798)[56], which states that if one factor of production is increased (in our case the writing) while the others remain constant (the data and analysis), the overall returns will decrease relatively after a certain point. It is a matter of training and skill to find the optimal point where to stop the adjustments in your writing.

Undetected typos are embarrassing, but they are disastrous if they change the content of your message. As an example, I quote from the paper that reports the main chapter of my own PhD manuscript, Sterken (1977): *"... similar short-term variations such as those seen in HD 160529 and eventually in HD 168607 have <u>never</u> been found in α Cygni"*, which states the opposite of the fact expressed in the preceding sentence. What happened? The submitted typescript correctly stated that such variations have **ever** been

---

[55]*Sparring partner*: any person with whom a professional boxer practices while training.
[56]Thomas Robert Malthus (1766–1834) was a British economist and demographer.

seen, but the typesetter[57] made a typographical mistake, and the proofreader (myself) did not spot the error, and gave his *imprimatur*.[58] The message can somehow be correctly understood from the context in the same paragraph, but citations out of context will convey an unintended opposite view.

Let me conclude this Section with an example from my editorial practice. I recently completed a conference Proceedings Volume together with two co-editors (Sterken *et al.* 2010). Both colleagues were, like me, non-native English speakers, and we set up a system of proofreading in three passes: I would first correct incoming manuscripts for typos, technical mistakes and obvious syntax errors. Then my colleagues would annotate the corrected manuscripts with their own remarks, and send them back to me. The experience showed that in about two thirds of the cases, each of us would spot the most evident language errors, and suggest a same correction. But in the remaining cases, it quite often happened that we had two or even three different interpretations of "what the author possibly wanted to say", and we then corrected the text in the most logic direction, knowing very well that a fourth party – the reader – may perhaps come to even another understanding. Do not forget that a similar situation may occur with the language editor, who in general is not a trained scientist, see Section 13.21 (and Sect. 10.2 in Paper III) for more details.

10.2 Etiquette and netiquette

Etiquette is a set of rules of conduct to be observed in social or official life, and *netiquette* is the etiquette of cyberspace. Whereas etiquette, as a code of conduct, may appear obsolete these days, business companies offering almost identical products at cutting-edge prices, find out that good manners may make the difference. The same is true when dealing with referees and Editors: sloppy conduct and bad manners may harm your cause.

The first rule is: always respond or reply to Editors and referees, also when your paper is being refused. Even sending a simple message acknowledging their mail (for example saying that you will revise the manuscript and submit again) is much better than not replying at all. By all means, avoid being rude: remember that your words are written, and that they can come back to haunt you.

The second rule is to respect the Editor's time and bandwidth by avoiding voluminous mails and unnecessary attachments. I had one case where an author sent me the inutile `.dvi` file, returned the `.sty` file, and even included Windows system files with the submitted archive. And, never submit a LaTeX file that you cannot compile, never tell the Editor that *"it correctly compiles after you hit* `return` *when it halts at an error message"*, because no Publisher will ever accept such a thing. Nor use home-made macros or local typesetting packages that are not available at the commercial printer's office.

Referees and Editors are working scientists, and have their own research deadlines and teaching duties. It is a matter of education and courtesy to submit to these people manuscripts that have received the highest degree of author attention as possible. It

---

[57] In those days, manuscripts were outsourced and sent overseas for typesetting.

[58] *Imprimatur* (from Latin) means consent to print.

does, of course, happen that referees let pass a poor paper. But you should by no means draw wrong conclusions when seeing such a paper in a high-impact journal: mirror your manuscript to the best papers in the journal, and do not submit a poor product because another author has succeeded in getting something published that is below standard.[59]

When a referee's report contributes to a significant improvement of the manuscript, or when the referee flagged a capital mistake, it is good practice to thank the referee (see also the last paragraph of Sect. 13.1, and Sect. 13.10). But do not thank the anonymous referee for *his* help unless you are sure that the referee is male (see Sect. 11.7).

11   What to avoid at all price

11.1   Writing things that you do not understand

Never add sentences that you do not fully understand (one of the pitfalls of the *copy and paste* facility). This is especially true for Latin abbreviations, like *et al.* that stands for the Latin *et alii*, and the very confusing *i.e.* and *e.g.*: the former stands for *id est* ("that is", or "in other words") whereas the latter means *exempli gratia*, "for example". There also is *viz.* – an abbreviation of the contraction from *vidēre licet* – which means "namely", used for clarifying something, and often is accompanied by a colon when the following statements are in a list.

Foreign terms (*Gedankenexperiment*) and other Latin words (*minus, versus, in extenso, ad hoc, a fortiori, etcetera, caveat,* ...) should, in principle, be written in italics. Some Publishers adopt the editorial convention of not italicizing the phrase "et al.".

11.2   Drowning in acronyms

Science uses a technical terminology that allows for significant gain in communication efficiency by coding concepts through the use of acronyms. Space science jargon, in particular, is peppered with acronyms for space vehicles and parts thereof. Try to avoid inventing acronyms, and if you really feel that an acronym should be used, then first count the number of times the acronym appears in your text: the introduction of a new acronym does not pay off if it is only used two or three times. Make also sure that the acronym you propose has no political or obscene meaning, and when using an acronym for the first time, you also give its definition. On page 14 the case is mentioned of two quite common acronyms, but very confusing situations can occur, for example GCPD (*General Catalogue of Photometric Data*)[60] is known by movie adepts as the fictional *Gotham City Police Department.*[61] This example may seem absurd, but it is nonsensical **only** because we know the context.

Never use acronyms that you do not understand, for example some observers may not know what a PMT (Photomultiplier Tube) is, nevertheless the acronym is used. Do not add a genitive to acronyms: "WET's results" is better written as "the results obtained by

---

[59] See the remark on verification and validation in Section 4 of Paper III.

[60] http://obswww.unige.ch/gcpd/gcpd.html

[61] Gotham City in the Batman books published by DC Comics.

WET".[62] As said already, no acronyms should appear in the abstract. Scientific terminology is unavoidable, but unnecessary jargon and complicated non-scientific terminology only slow down the reading process.

11.3   Mixing American & British styles

A paper or a book should use either American or British style. British words ending on *-our, ise, -logue, etc.* are spelled differently in American (color, analyze, catalog, etc.) – but look out for the lack of logics, see the last two lines in Table 1 (*i.e.* devise, revise, scrutinize, utilize). Note also that the Americans write *acknowledgment* for *acknowledgement*, and *cesium* for *caesium*. See also the paper by Adams *et al.* in Part 1 of these Proceedings, and a very useful list of accepted spellings (following government guidelines and major dictionaries) for UK & Ireland, South Africa, New Zealand, Australia, Canada and the US compiled by WIKIPEDIA.[63]

11.4   Clichés

A *cliché* is something that has become a commonplace. The term *asteroseismology*, for example, has become such an attractive term to use, that it appears in so many introductions to papers that have very little to do with the seismology of stars. Quite often an author has determined only a couple of frequencies – not even a complete frequency spectrum – but still wants to sell the product by opening the paper with the magic word: there is just no need for such marketing using a product label that does not cover the content.

11.5   Inconsistent capitalization

Some words may have a different meaning when capitalized: the Moon is our moon, whereas moon refers to any natural satellite, and Galaxy refers to our galaxy. A figure can be a printed character or a numerical value, whereas a Figure is a graphical representation. A table is a piece of furniture, a Table is a systematic arrangement of data in rows and columns. A Section is a group of paragraphs, a section could be a part or division, a team, or an incision, or a term from geometry – such as volume, whereas Volume refers to division of a book or a journal. In this paper I capitalize Editor and Publisher when I refer to the Editor or the Publisher[64] of a book or a journal, but when not capitalized, the terms refer to editors and publishers at large. Adjectives (like "editorial") are never capitalized.

11.6   Footnotes and unnecessary emphasis

Footnotes interrupt the reading process, and render a very poor page layout in Sections with lots of mathematical formulae. Footnotes frequently occur in review papers, and

---

[62] *Whole Earth Telescope.*
[63] http://en.wikipedia.org/wiki/Wikipedia:Manual_of_Style
[64] Capitalization, in this case, has also become a symbol of respect.

the full-page layout is better adapted to footnoting than does the two-column layout in a research journal. This book contains numerous footnotes, in particular for the many URL addresses. Keeping consistency in punctuation and capitalization in footnotes is not always evident: in Papers I–III, footnotes containing a complete sentence are closed with a period, except when the last item in the sentence is a web address. Short explanatory statements are not concluded with a period, nor do they start with a majuscule.

Some authors too often use boldface or italics to give some words and sentences particular prominence. Over-emphasizing simply works the opposite way – though for tutoring (like in this book), over-emphasis is often done on purpose. Pay attention not to use superfluous words, for example in "existing calibrations": there is no point in emphasizing "existing" since non-existing calibrations simply do not exist.

11.7 Sexism

Sexism is the collection of attitudes that foster stereotypes of social roles based on sex. Traditionally, the third personal masculine pronoun "his" was used to refer to masculine and dual-gender nouns (like astronomer). In modern usage this is considered sexist and therefore "his/her" is sometimes recommended to avoid sexism. But this is awkward, especially when used repeatedly in a text, it draws attention away from the real message, and also overloads the parser. One solution is to make the sentence plural. While it is possible to draft a paper and avoid a single occurrence of such sexist elements of expression, one should be careful when reading older papers and texts from times when such concerns were not raised. In my papers I have tried to avoid the problem, but it was not always possible to unwind any sex-based differences in communication style. But all references to gender in given examples or cases, refer to genuine situations.

Take the following example. When queuing at a Brussels-airport security gate, I saw a huge cartoon with the quote *By believing in his dreams, a man turns them into reality*, which is an homage from Hergé[65] to Neil Armstrong and the Apollo 11 crew. Struck by what would now be seen as sexist language, I consulted the original French text: *A force de croire en ses rêves, l'homme en fait une réalité.* Here, *l'homme* – man – was poorly translated to "a man". This vividly illustrates the danger of inadvertent misquoting in translation, see also page 18 and Section 11.9. And this, again, demonstrates that writing is a process: I must have spotted this phrase dozens of times without noticing the point I am making now. So why, exactly on my way to that particular meeting, was I struck by this imperfection? The answer simply is: because I was composing my paper in the weeks preceding my travel – even during my trip to the meeting covered in Sterken & Aerts (2006).

11.8 Non-standard nomenclature

When previously-defined designations of celestial objects are used in your paper, they should never be altered (*e.g.*, neither truncated nor shortened). Do not invent your private

---

[65]Pseudonym of Belgian artist Georges Remi (1907–1983), spiritual father of *Tintin*.

celestial-object name, while referring to the original consolidated nomenclature as "other designation": this creates confusion when other authors take over your naming scheme. Be honest and do not say that you discovered 23 new variable stars when half a dozen of them were already known and labeled before you started the study, and you simply renamed them! The original bibliographical reference for a designation should be given in the first place. Laurent Cambrésy, in his paper on archives (in Part 1 of these Proceedings), stresses the importance of this guideline, and I like to refer to a personal communication by Dubois (1995), a colleague of Cambrésy at CDS:

> CONGRATULATIONS ON YOUR PREOCCUPATION WITH THE PROBLEM OF DESIGNATIONS. IT IS A MATTER THAT IS TOO OFTEN IGNORED AND THAT CAUSES A LOT OF PROBLEMS. MY REACTION IS THE SAME AS YOURS: IF THE STARS HAVE ALREADY BEEN NAMED, IT IS PREFERABLE TO USE THE ARCHIVAL NAME . . . [66]

11.9   Misquoting, misciting and quoting out of context

Misquoting and misciting occur at several levels, and most incorrect quoting and citing are due to inaccurate reading and sloppiness. The following common errors occur:

- simple typographical errors in any of the fields listed in Section 9.12;

- quoting or citing without reading, or even having seen the source paper;

- verbatim copying references from a list without checking the source paper;

- misquotation by changing content;

- not citing at all, which definitely is the worst form of misciting;

- changing the order of names when citing a reference, in particular upgrading the position of your own name, and

- the (rarely occurring) omission of references in order to avert hampering the author.

The last point refers to tacit omission of faulty papers, and only once I found a direct reference to such an act in Bottle (1973): *"Reference omitted to avoid embarrassing its author"*, an action that of course results in just the opposite effect. Deliberate misquoting is considered as scientific misconduct, and is discussed in Section 7.8 of Paper III.

Typos are a nuisance, of course, but confusing abbreviations for journal titles are also to be avoided. For example, "Ann. Phys." stands for *Annales de Physique*,[67] but also for *Annalen der Physik*.[68] The A&A instructions list three journals starting with A&A in Appendix D.[69] Another point of confusion, now becoming extinct, is the habit of

---

[66] *Félicitations pour vous préoccuper du problème des désignations. C'est une question trop souvent ignorée et qui provoque bien des difficultés par la suite. J'ai la même reaction que vous : si les étoiles ont déjà eté citées il est bien préférable d'utiliser les dénominations antérieures . . .*

[67] One of the EDP Sciences journals.

[68] Wiley InterScience.

[69] Simplified abbreviations of frequently used journals (http://www.aanda.org/images/stories/doc/aaguide_v7.0_201001.pdf).

replacing the name of the author or book by the Latin expression *ibid* or *op. cit.*[70] These abbreviations should only be used when a reference source is referred to several times in a short space – remember, memorizing several ibidems will contribute to overloading the reading parser.

A frequent error occurs as a consequence of downloading, when the correct URL is quoted, but when the reference to journal and page number is wrong, or is omitted. For example, Sterken (2006) is cited by Stein (2006) with a correct URL, but the paper is not attributed to its author, but to the two Editors of the Proceedings book (*i.e.* Sterken & Aerts 2006), and Stein (2006) fails to give the correct reference. The principal consequence is that the erroneous citation is copied into another paper without the author/copier ever having seen or read the original document, hence perpetuating the error in a cascade of papers. This, unavoidably, biases bibliometrics (see Sect. 18 of Paper III).

The most famous example of citing without having seen the original – over more than half a century – involves a paper in the medical literature entitled *O úplavici; Předběžné sdělení*[71] written by the Czechoslovak author Jaroslav Hlava in 1887. In the same year, the *Zentralblatt für Bakteriologie und Parasitenkunde* states[72] "Uplavici makes the interesting communication. . .", as such creating a fictitious author "Uplavici, O." and truncating the article title to its second part *Předběžné sdělení*. The mistake led to numerous citations of the work of "Dr. Dysentery", and after exactly 50 years the situation was remediated by Dobell (1939), who lists all varieties of this author's name: O. Hlava, Hlava Uplavici, and even the two "persons" Hlava and Uplavici.

Misquotations can survive for centuries, take for example *e pur si move*: these are the words which Galileo Galilei is said to have uttered on the 22nd of June 1633, after abjuring the heliocentric doctrine. The origin of this legend is hard to trace, and many variants of the expression are found in the literature: *e pur si muove*,[73] *eppure si muove, eppur si move, eppur si muove*, and so on. Berthold (1897) traces the first written version "eppur si move" to a publication in 1757 by Giuseppe Baretti (1719–1789), but there is no solid trace to the original statement, if it was uttered at all. As Galileo was Pisan, and the Italian language – in particular the literary language – was basically a form of Tuscan, it was not unlikely that the words reportedly uttered may actually have been *"eppur si muove"*. Galileo himself, in his essays, was a very good stylist and in any case the distance between his written language and his spoken was unlikely to be great (Lepschy 2006). Galileo's alleged uttering is most probably as much of a fiction as is the fable about the EUREKA of Archimedes.

Another source of miscites are the so-called textbook errors, which often are repeated with little or no opportunity for correction. One example, pointed out by Wróblewski (1985), is the incorrect description of Ole Rømer's work on the velocity of light in many physics texts and books dealing with the history of science. Figure 3 illustrates the extent of this misinformation: though Rømer never gave any numerical value for the speed of light, dozens of authors attribute a numerical value to him. About half of the values hover

---

[70] *ibidem*: in the same place; *opus citatum*: the work cited.

[71] *On dysentery; a preliminary communication.*

[72] "Uplavici macht nun jetzt die interessante Mittheilung. . ." 1887, Jena verlag, p. 538.

[73] And yet it moves.

around $220\,000$ km s^{-1}, the remaining ones are in the range $300\,000$–$350\,000$ km s^{-1} (see Paper II for further discussions on this bimodal distribution), and this is due to the calculation being extremely sensitive to the value used for the Astronomical Unit, a quantity that was quite poorly known in Rømer's times.

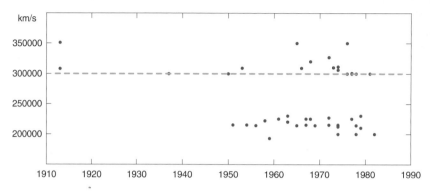

Fig. 3. Numerical value for the speed of light quoted in history of science and physics texts and books as "derived by Rømer". The dashed line represents the physical constant c. Source: Sterken (2007).

Misquoting by changing content occurs when the original content is modified, truncated or cited out of context. Note that even changing the font in part of a sentence can change the meaning. If you really wish to quote, and also emphasize part of the quotation, you must specify (for example in a footnote) "italics mine" or something similar.

A very special case of misquoting by changing content is when the original quote comprises an error or misspelling: in such a case one should not merely correct the error, but reproduce the passage *verbatim*, and insert the Latin word *sic* in square brackets right after the error to which it refers. *Sic* means "as such", and helps the reader's parser out of the confusion about who is wrong: the quoter or the quoted.

Mind, though, that the application domain is much wider than typographical errors. Take, for example, the following sentence from *Lady Chatterley's Lover* (D. H. Lawrence's controversial novel written in 1928): *"Why is the star Jupiter bigger than the star Neptune?"* From an astronomer's viewpoint, and out of context, this statement is wrong, since it refers to planets. But from the author's perspective it is not necessary inappropriate, since the sentence is uttered by Sir Clifford Chatterley, and obviously refers to the celestial lights Jupiter and Neptune, where "bigger" certainly denotes perceived brightness. In addition, the mindset of the present reader is totally different from the one of the writer nearly one century ago. Here, *sic* would be of good use, since quoting out of historical context also is a mode of misquoting. Overuse of *sic* should be avoided: for example, on page 25 the spelling of "Colours" is different from the way it appears in the rest of this book, yet this divergence needed not to be flagged with *sic*.

Inaccurate quotations and citations, whether intentional or not, simply increase the noise in the information transmission system (see Sect. 1 in Paper II), and cheat the citation rankings. This also occurs when citing obsolete papers: avoid citing or attacking positions that are no longer defended.

12   Instructions and guidelines

This Section deals with instructions, and focusses specifically on submitting a manuscript to an Editor of a conference proceedings book, but some of these instructions and guidelines are also useful to authors of journal research papers.

There are two sets of guidelines: Publisher instructions, and Editor instructions. The first set of rules is described in detail in the Publisher's *Instructions for Authors*[74] that define the "house style" of a journal. Respecting these rules is just mandatory, for journals as well as for Proceedings books.

12.1   Editorial instructions

Editorial instructions are normally provided by the Editor(s) of a book. One of the major mishaps when editing books or conference Proceedings occurs because of authors' poor understanding of the system of references, in particular the thebibliography environment in LATEX. Though the use of this environment allows most versatile handling of citations and references, many authors do not realize that formatting errors in the reference list can propagate through a complete book and, consequently, destroy the coherence of the bibliographic citations and pagecites. Even a manuscript that correctly compiles as a stand-alone paper, may render another paper's citations chaotic and useless. A second nuisance is when an author does not respect margins, especially with overflowing running titles, URL addresses, Figures, Tables, and their captions.

Editors of a book or conference proceedings must practice very systematic and tight working procedures.

- Editors frequently deal with over a hundred LATEX files, and perhaps several hundred graphics files, when compiling a single book, hence authors are advised to name their LATEX and encapsulated postscript files in a systematic way, such as surname.tex, surname-fig1.eps, surname-fig2.eps, etc. For example, for Sterken & Aerts (2006), I received several firstname.tex files – an impossible task considering the duplication of half a dozen first names in the audience. It is also not uncommon that an Editor covering a meeting in, say, Brussels, receives multiple files brussels.tex, fig1.eps, fig2.eps, etc.: in such cases all these files have to be renamed in order to create a workable and efficient compilation environment, and this process takes time and leads to errors.

- The confusion between ENCAPSULATED POSTSCRIPT (.eps) and POSTSCRIPT (.ps) formats, which are standard (ASCII) formats for importing and exporting POSTSCRIPT language files in all environments (Section 5.4 in Paper II). The file must include a header comment starting with %!PS-Adobe-3.0 EPSF..., and a %%BoundingBox comment that describes the bounds of the illustration. The .eps file can contain any combination of text, graphs, and images. An .eps file is the same as any other POSTSCRIPT language page description, with some restrictions, and is usually a single page. Some authors label their Figure files with the .ps extension. While not

---

[74]For A&A, see the author's guide on http://www.aanda.org/content/view/136/184/lang,en/

basically wrong, this habit creates confusion, as the `.ps` extension is generally used for multipage documents (a complete paper or book). Mixing `.eps` and `.ps` extensions for one-page documents to be inserted in a TEX manuscript certainly does not help journal and production Editors keeping a tidy job.

- If page limits are imposed, they should be respected. In practice, allowing even a one-line overshot unavoidably results in two excess pages, as papers quite often start on odd pages. Authors should be aware that also Editors have page limits, in the sense that extra pages must be paid for from their budgets.

- Some authors submit subsequent versions of their LaTeX file after their first submission already underwent considerable editorial labor. It is just not possible to track down such changes, nor is sending back a source file to the author a workable option, since the Editor may lose all control over the changes made previously. Moreover, Editors often simultaneously modify multiple TEX files for adding index entries. It requires the editorial experience of just one book to know that not every author understands how to efficaciously contribute to the editorial process.

- Late (or no) submission is one of the most irritant problems, as it has a direct impact on the editorial planning. Some excuses are very common: disk crash, laptop stolen, etc. The worst justification for a really long delay I ever received was *"I am waiting for the permission to include one more illustration"*. The most pleasant statement I welcomed was *"I apologize for the delay in sending the files (for which there is no real excuse)"*. Proceedings book editing is always done along with the complete load of other work like observing, teaching, data analysis, meetings, vacation, family business, etc., and a delay of a couple of weeks can make a hell of a difference in an Editor's planning. And if the manuscript is never submitted, there is not much an Editor can do, except boldly publish the title of the talk with a statement in the way Schielicke (1998) did:

  > TO THE REGRET OF THE MEMBERS OF THE BOARD OF THE ASTRONOMISCHE GESELLSCHAFT, OUR 1997 AWARDEE OF THE KARL SCHWARZSCHILD MEDAL WAS NOT IN A POSITION TO PREPARE THE MANUSCRIPT OF HIS LECTURE, AS WAS AGREED BEFORE.

Let it be clear that Publisher instructions always override Editor's guidelines, and that these conventions are not just a Publisher craze to take the fun out of your work, but are given to ensure cost-effective streamlining of the publishing process towards a coherent and handsome publication in the house style of the Publisher. When I was editing a conference Proceedings book, one author wrote me that he refuses to use the `\section` command, because it spoils space, and he replaced the command by one of his own making, rendering his paper different from all others in the book. Another author just submitted an MSword file including mathematics, expecting that the Editor would do the homework. Needless to say, both approaches were totally unacceptable.

Publishing has changed dramatically in the past decade: simply compare a conference book of the 1980s with a modern one. Many of the older books have a mixture of typefaces and fonts,[75] variable line spacing and variable margins, and a most heterogeneous collection of reference formats. Modern Proceedings books are aesthetically more pleasing, and more efficiently organized, in the first place because of tight instructions.

After all, accurate rules also prevail in electronic banking (accuracy of amounts and accounts), observing proposals (coherence and deadline), grant applications (deadline, structure and logic), even placing an order for your favorite music at a webstore implies abiding to strict procedures.

12.2 Copyrights

Copyright deals with the exclusive legal right to reproduce in publications, and is usually transferred to the Publisher through a *Copyright Assignment*. Most scientific authors – young and old – just blindly sign such a copyright-transfer contract, and many a participant is rather surprised when confronted with the content of such a form, which is mostly quite cleverly worded. Take as an example the following paragraph in the Copyright Assignment from the *Astronomical Society of the Pacific* (ASP) addressed to you, the author:

> ... BECAUSE YOU DESIRE TO HAVE THIS PAPER SO PUBLISHED, YOU GRANT AND ASSIGN THE ENTIRE COPYRIGHT FOR THIS PAPER EXCLUSIVELY TO THE ASP. THE COPYRIGHT CONSISTS OF ALL RIGHTS PROTECTED BY THE ORIGINAL COPYRIGHT LAWS OF THE US AND OF ALL FOREIGN COUNTRIES, IN ALL LANGUAGES AND FORMS OF COMMUNICATION.

This means that you transfer your entire copyright to the Publisher. In this particular case, you retain the permission to reproduce your work elsewhere, provided you give appropriate credit. Note that some copyright assignments leave you with much less liberty than this one. Beware that you cannot legally sign two Copyright Assignments and submit the same manuscript to two copyright-demanding Publishers, see Section 7.12 in Paper III.

Almost all book Publishers take away your copyrights, and very few journals leave copyrights to you (the copyright holder for the papers published in A&A is not the journal, but the European Southern Observatory, ESO).

Evidently, the financial consequences augment with the commercial value of the material that you copy: just be prepared to fight lawyers if you breach copyrights with a well-established Publisher, museum, or library. Let me give you one practical example. In Sterken (2005), I reproduced three graphics: a color reproduction of two portrait paintings on the cover, and one black-and-white reproduction of a comic strip inside the book. Permission to reproduce the portrait of Ole Rømer was granted free of charge by the Rundetårn Museum in Copenhagen: hence I reproduced it inside the book (in black-and-white), and also on the cover of the book (in color). Permission to reproduce the portrait of Giovanni Domenico Cassini was granted by Observatoire de Paris for a fee of less than 50 Euro per reproduction, hence I reproduced it on the cover only. Permission for a single black-and-white reproduction of a *Tintin* comic strip was granted for an almost ten times higher fee, and several well-specified restrictions were imposed (no modifications

---

[75] Typeface and font are explained in Section 7.1 of Paper II.

of text balloons, for example). Likewise, the permission to reproduce the photograph of the tapestry on the front cover of this book, was purchased from the *Röhsska Museum of Fashion, Design and Decorative Arts* (Göteborg, Sweden) by Publisher EDP.

13   FAQs about the editorial process

13.1   Who should be coauthor?

Appropriate authorship assignment is of great significance and value to the scientist, because it contributes to the building of reputation, to the assignment of responsibility, and to recognition of achievement. In addition, correct authorship is crucial for the scientific grant administration, because authorship and citation are observables that contribute to the scientific rating of a project or of a person.

It is well-known that senior scientists do influence decisions on authorship, and that some senior scientists sign for authorship at the expense of junior researchers. And it is also known that senior authors have moved to first authorship at the cost of others, that authorship is granted to chairpersons of departments as a matter of convention, or that senior authors are listed just to boost the paper, see Drenth (1998) and references therein, and also Sections 7.3–7.5 in Paper III.

The *Vancouver guidelines*[76] attempt to define the contributions that allow coauthorship, and suggest that authorship should be based **only** on substantial contributions to conception and design – or analysis and interpretation – of data, **and** profound involvement in the publishing process. A very important element – for us, astronomers – is subsequently formulated in the so-called *Copenhagen compromise*,[77] which adds that **the acquisition of data merits authorship – provided that those involved in the acquisition of data also meet the other criteria for authorship**. So, the guideline is that authorship credit should be based solely on substantial contributions to:

> 1. CONCEPTION AND DESIGN, OR ACQUISITION OF DATA, OR
>   ANALYSIS AND INTERPRETATION OF DATA,
>
> 2. DRAFTING THE ARTICLE OR REVISING IT CRITICALLY
>   FOR IMPORTANT INTELLECTUAL CONTENT, AND
>
> 3. FINAL APPROVAL OF THE VERSION TO BE PUBLISHED.

All three conditions must be met.

These guidelines reject fund raising, and general supervision of the research group as justifications for authorship, and also the bare collection of raw data – though these contributions should be acknowledged. The guidelines do, however, make it clear that between them the authors must take responsibility for all aspects of the work: **any part of an article critical to its main conclusions must be the responsibility of at least one**

---

[76]The *Vancouver Group* refers to the *International Committee of Medical Journal Editors (ICMJE)*.
[77]ICMJE, May 2000 meeting, http://www.icmje.org/urm_main.html

author. This brings up the question of how to correctly evaluate papers coauthored by literally dozens – if not hundreds – of authors, especially in the event of large projects that manage tight deadlines. In such cases it is, alas, quite common practice to enwrap a complete team in authorship, with barely enough time for each envelope member to critically respond to draft versions of the manuscript. Such practices not infrequently lead to complications, see Paper III.

The *Vancouver guidelines* are, in fact, already implied in service-observing contributions, as major observatories request an acknowledgment of the service-observing share of the work (for example, phrased as *"the authors thank N for doing a wonderful job in organizing the service observations"*). But in practice it does happen that service observers only hear *a posteriori* that a manuscript with their name on has been submitted, so they do not even get a chance to respond nor to contribute to the writing.

A question I am often asked is whether the use of a computer code written by someone else is a contribution that merits authorship. I see such a case as an analog to service observing: if the code was not specially written for a specific paper, and if the writer of the computer code does not participate in any of the other aspects of the analysis and the writing of the paper, then that person should not be a coauthor, but should be mentioned in the acknowledgments instead.

Both types of examples imply no involvement in analysis or interpretation, nor in writing the paper: they thus fail to meet any criterion of responsibility or accountability. Many a service observer or computer programmer, who involves as coauthor without participation in the data analysis or in the writing process, has later on been confronted with the results of inappropriate handling of the raw data, or with improper application of a computer program.

In summary, most scientific papers have several contributors. Some of these people are to be listed as authors, provided that some requirements are met. The remaining contributors (language editors, referees, supervisors, service observers, data archivists, librarians, ...) are to be mentioned in the acknowledgments paragraph at the end of the paper, see also Section 13.10.

13.2   What is honorary authorship?

Honorary authorship is authorship given to colleagues who did not contribute to the research or to the communication of the results, and is akin to *gift authorship*: granting authorship to people who did not make any intellectual effort. This kind of present is often mechanically assigned to senior colleagues since long retired, but the mechanism is also applied for profit by borrowing the name of a celebrity. Honorary and gift authorships not only are demotivating for those who did the real work, such arrangements can be quite risky for the recipients. Section 13.3 states that the first author, in general, bears the responsibility for the complete work. But if there are senior honorary coauthors in the list, and if the paper is rejected because it does not fit the scope of the journal, then the first-author responsibility rule leads to a paradox, since the seniors should have stopped the submission to an unsuitable journal in the first place.

Avoiding inclusion of gratuitous coauthors, evidently, also applies to yourself on other people's papers: exclude yourself if you have not contributed to a paper for which you are invited to sign as coauthor – although it must be recognized that the rising trend of papers with dozens of authors does not make such decision easy. Small teams do sometimes work with a kind of reciprocity, *i.e.*, mutual exchange of participation in each other's papers. This is a habit to avoid because you may end up disappointed (reciprocity is not always guaranteed and is often forgotten), and you may even be blamed for grave errors. For more details on related issues in the context of multisite observing campaigns, I refer to Sterken (1988).

13.3   How to settle the order of author names?

In some fields of science, authors are always listed in alphabetic order,[78] but in astronomy author names are mostly arranged in order of decreasing contribution to the work. A solution that I applied during quite some time in my early postdoc years, was to decide the author sequence at the very end of the writing phase, just before submission. But that ideal does not fit every team, and turned out to work only when no more than two or three authors were involved. Even two-author teams have shown to run in trouble when it comes to deciding who comes first.

Remember that the first author, in general, bears responsibility for the work. The perceived importance of the last author is more difficult to evaluate: it can be seen as "the last is the least", or that the last name is the supervisor (with possibly the same negative appreciation), or that the last author can be an important senior contributor who insists on that position.

The main point is that the code used by the authors in constructing the hierarchical order is only known to the authors, and not available to the reader or Editor (in fact, the code is relatively undefined). Therefore, more and more journals (especially in biomedical sciences, but also *Nature*) require or encourage the authors to describe their specific contributions to the paper they submit. The acknowledgments paragraph at the end of the paper is the ideal place to explain what each author has contributed.

Sheskin (2006) proposes the *analytic hierarchy process*, a method to enable coauthors to determine their fractional contributions to a multi-authored paper. The procedure consists of dividing the paper into primary and secondary elements – for example, "Methods" in the IMRAD scheme would be a primary element, comprising light curves, frequency analysis, spectra, ... as secondary elements. Each author's contribution to each part can then be readily assessed, and the relative merit can be calculated. The implementation of this method remains to be seen.

One should always discuss authorship and author ranking in an open way: if you think you should be the first author, then start a dialogue with the others, and give your arguments. **Never** change author sequence on your own initiative only: dialogue is the only solution if you do not wish to permanently damage your relationship with colleagues with whom you collaborate (see the case discussed in Sect. 8.7 of Paper III).

---

[78] In mathematics, for example.

13.4   What if a coauthor passes away?

It does happen that a coauthor – even the first author – dies before a manuscript is composed or submitted. If that person has contributed in a substantial way to the project, then one should consider including that person in the list of authors. The proper way to do this is to mention the complete affiliation of the place where the deceased was working while committed to the project, and add in a footnote a statement expressing that this coauthor has passed away on a specific date.

13.5   Can the list of authors change during the review process?

Yes, it happens, but it should be avoided. It can occur that a listed author asks to be removed, but such an action mostly points towards shortcomings during the preparation phase of the manuscript. There are few good reasons to add names after submission, for instance it can happen that a referee contributes so much to the content of a paper, that the author(s) may decide to offer coauthorship to the referee. Extending coauthorship to the Editor of a journal, however, is considered as inappropriate (Hames 2011).

13.6   Who submits the manuscript to the Editor?

The *Corresponding Author* is the person who submits the paper for publication, and who is responsible for the manuscript as it moves through the editorial line. This person, most often, is the first author, and manages all correspondence with the editorial office, and coordinates the communication with the referee(s) through the manuscript management system, using the author identification provided by the journal. In addition, this author releases the proofs, and deals with the administrative aspect of organizing the payment of page charges. In my opinion, it is crucial that a PhD student – if that person is first author – also becomes corresponding author.

The BMJ Group, who publishes *The British Medical Journal*, proposed the concept of *guarantor*: the author who is prepared to take responsibility for the whole article. This system is not used in any of the astronomy journals I know of.

My experience as an author and as a referee is that quite a number of journal Editors do not engage in direct contact with their authors and referees, and most communication is taken care of by an editorial assistant,[79] at least as long as there are no disputes or conflicts. I must say that, as a referee, I have always enjoyed the direct contacts with Editors of journals who still nurture the principle of direct communication with their referees and authors.

13.7   Must I wait for a coauthor's consent to submit?

Yes, you must wait till all authors have read and agreed on the last version before it can be submitted. Likewise, coauthors have the obligation to verify and validate not only the

---

[79]Editorial assistants, in most cases, are not scientists.

submitted manuscript, but also the revised version. I illustrate the hassle with three cases in which I got involved.

1. In the case of a two-author manuscript where I was second, one of my students submitted to a journal without telling me, nor even showing me the last version of the manuscript. The referee required that the paper be completely rewritten, which led to a major delay and even withdrawal in the end.

2. One paper, with six authors, spent many months in the reviewing process, with multiple passes through the referee's hands. The final version had undergone a major facelift, including corrections and removal of erroneous parts, so I requested the first author to contact every coauthor in order to obtain their consent before submitting the final version. Four of the coauthors had not followed the evolution of the manuscript during its revisions – which included fundamental changes – yet they merely consented without any discussion whatsoever. One coauthor never even replied.

3. As Editor of a Proceedings book, I once received a paper signed by three authors. Not only was the page limit transgressed, the syntax also revealed that the native-English speaking coauthors could not possibly have seen the submitted manuscript. So, I copied my reply to all coauthors, asking to shorten the paper and to language-edit the manuscript. The second author complied, and submitted a revised version, after which the first author requested that the changes be undone. This way of working is most inefficient, steals time from all parties, and spoils the professional relationship between all persons involved.

You may not realize how easy it is for the reader (and referee and Editor) to tell such state of things apart. Submitting a manuscript before all coauthors have approved is bound to lead to problems, creates a very embarrassing situation for some coauthors, and may even produce the vision that the supervisor does not even bother to read the students' manuscripts before they are submitted. **Journals require that the submitter implicitly or explicitly declares that all authors have seen and agree on the manuscript that is submitted,** so when one or more authors do not confirm their consent, the manuscript should simply not be submitted.

The *Vancouver guidelines* clearly state that drafting the article or revising it, is a basic **duty** of every coauthor. But the **right** to at least see, read and approve any submitted version is a fundamental right that should not be denied to anyone on the author list, see also Section 13.2. In my editorial practice I am still astounded to see, again and again, that so many senior researchers seem not to know this rule.

13.8   When to submit a manuscript to astro-ph?

The idea to submit your manuscript to a journal, and at the same time to deposit it at astro-ph, is not widely recommended for the following reasons:

1. the deposited material is not peer-reviewed;

2. the file is archived with reference to the journal to which it was submitted, so in case that the paper is refused and submitted to another journal, the reference will be incorrect;[80]

3. if the peer review reveals gross errors, then the flawed manuscript may remain available to the community at large. If the revision process takes considerable time, it is not impossible that someone reacts with a counter paper even before your paper is finalized.

Let me illustrate the problem with a real example. An author submits a manuscript to a journal, and at the same time deposits it at `astro-ph`. But the referee disagrees with some of the conclusions, and substantial parts of the paper have to be removed. The paper is finally accepted, but the total revision time (with two passes through the referee's hands) takes a full 11 months, and the acceptance year is one calendar year later. In this particular case, the `arXiv e-print` today still renders the originally submitted flawed version, whereas ADS does lead to the correct abstract and bibliographic code, but with the link pointing to the 1-year older version available at `astro-ph` (see also Sect. 3.2.1 in Uta Grothkopf's paper in Part 1 of these Proceedings). Bibliometric bias is not unlikely to develop when only the `astro-ph` server is consulted.

In fact, the situation described above also occurs when presenting results in poster format at a meeting, if the poster is based on a submitted but not yet accepted journal paper. If the referee finds the main paper unacceptable for publication, then the poster paper should be withdrawn of course. This, obviously, seldomly happens.

13.9   How to communicate with the referee?

Communication with the referee is always done via the Editorial Office, and not directly with the referee. This interaction is carried out by the corresponding author. The four main categories of responses by a referee are:

1. acceptable with minor revision;

2. acceptable after major revision to be verified by the referee;

3. outright rejection – mostly for scientific content, but also for poor language, inadequate graphics, and badly structured or incoherent narrative, and

4. rejection because the paper does not fit the scope of the journal.

The last outcome is always somewhat regrettable, because it is a result that could have been foreseen[81] by the senior coauthors. There are two ways of pursuit here: either submit to another journal, or oppose the Editor's decision (for example, by providing a precedent case of a similar article that was accepted for publication in the same journal). In the latter case, it is by and large better to let the senior coauthors do the fighting.

The third scenario is the most painful to undergo, although the second category can also be quite difficult to deal with, because soft-looking and polite statements by a referee

---

[80]This feature even is a source of information for Editors.

[81]For example, if you submitted a review paper to a journal that does not accept reviews.

may hide a lot of disagreement (in just the same way as a soft and positive letter of reference may not get you the job you are looking for).

Learn to correctly read the referee's message: sometimes the referee asks a question that looks stupid, at first sight. That does not necessarily mean that the referee is slow of mind: it is very probable that the referee, by playing devil's advocate, points to some confusing or vague elements in some passages. Few written statements are misunderstood because the reader lacks intelligence: most of the incorrect interpretations result from foggy writing (like using words and verbs from Table 1 in the wrong context). Referees – like lawyers – surely interpret ambiguous sentences in the way **they** think is just, not necessarily in the way the author thinks is correct.

Explain the changes that you have made to the manuscript (do not just silently expect the referee to find out what you have changed), and spell out why you prefer not to follow some of the referee's suggestions. If you plainly disagree with the referee, then open a dialogue and provide supporting evidence for your viewpoint – through the Editorial Office of course.

When receiving a very negative referee report, never answer immediately on line, but wait at least twelve hours before responding by email or letter. There is a fair chance that, reading the report over the next day, you may find it less shocking or to a lesser extent impossible to digest.

Be forgiving of other people's mistakes: also the referee makes typographical errors, especially because the reports are always due by a deadline, and the closer to the deadline, the less time is left for the referee to proofread his or her report. I once requested an author to add a reddening vector and error bars to their "Figure 5", but in fact I mistyped the Figure number, and meant instead to refer to their Figure 6. The author replied *"Figure 5 is a power spectrum and then how to add reddening vector and error bars?"* Such sarcasm does not boost a smooth and fast reviewing process, and is as inadequate as insulting your referee at a sports competition. On my suggestion that proofreading by a native English-speaking person would be very beneficial, the same authors replied *"Actually our manuscript was sent to a native English speaker for proofreading with charge. ...It would be appreciated if you would revise the manuscript."*

In fact, any form of ironic riposte is out of place in a response to the referee, not to speak of brutal replies which, regretfully, appear from time to time.

13.10   When to thank the referee?

Referees remain, very often, anonymous. Yet, sentences like *"The author thanks the anonymous referees for their valuable comments"* regularly occur in the literature. Kozak (2009) lists several considerations why authors are keen to acknowledge referees:

1. the authors think that the comments they receive are of such a high value that the referee's contribution must be acknowledged;

2. the authors think that it might help them to have the paper accepted because it makes the referee feel happy;

3. it is just that author's standard procedure;

4. they are asked to do this by the journal Editor.[82]

Kozak believes that the majority of referees are aware that they may be acknowledged with the second motive only, and that they may have become quite resistant to this kind of acknowledgment. Concerning the acknowledging-by-default of the referee, he mentions that there are even authors who are so eager to thank their referees that they do so in a first submission! Concerning thanking because the Editor of a journal asks to do so, he makes it clear that he does thank referees only if their comments are useful, a very correct attitude indeed.

In summary, it is required to thank the referee if that person merits it by having discovered major flaws, or by having made significant suggestions to improve the scientific quality of the paper. When the referee suggests only minor changes, an undeserved flattering acknowledgment may even be seen as an insult.

13.11   Can a student be referee?

Students may occasionally be asked to review a paper of a senior scientist. The first thing to consider is whether there is any conflict of interest – that is, when close friends or colleagues with whom you collaborate are on the author list. You should then normally refuse becoming the referee. Do not ruin given chances so early in your career, and never hope to hide behind anonymity: abuse of such powers will always backfire on its originator (see Sect. 8.7 in Paper III).

13.12   With whom can I share a referee report?

At the moment that you receive a referee report, there are only three individuals who know its content: the referee, the Editor, and yourself. The only other people that have the right (and the sole right) to read its content are the coauthors. You may decide to share the content with your supervisor, but a referee report is not intended to become public domain. Vice versa, if you become a referee, then your report should not be shared with anyone, not even for language editing. If you wish to consult other colleagues, you must first confer with the Editor in order to avoid involving other experts who may already have been excluded by the Editor or the authors.

A related question is if referees can identify themselves to authors: this can be considered – if the policy of the journal tolerates disclosure of the referee's identity – and it should only be done via the Editor. Vice versa, authors should not attempt to identify referees, and I refer to more discussions on this topic in Steve Shore's paper in Part 1 of these Proceedings.

13.13   May I reproduce complete sentences?

Several students asked me whether they must get permission to reproduce complete sentences. In fact no, because of the so-called *Principle of Fair Use*. This principle is a

---

[82]Or quite often by the thesis supervisor (point added by myself).

privilege that allows users to copy without permission for non-commercial purposes of criticism, research and teaching. Thus, **when citing one or a few sentences, provided you give due credit, no permission is needed**. But the principle is vague, and when in doubt, it is better to ask for permission, though you should realize that most journals are published in a commercial environment ("for profit"), and that you will likely have to pay a fee. The procedure is very simple: visit the website of the *Copyright Clearance Center*[83] (CCC), which offers easy-to-use licensing services, select the journal you wish to reproduce from, complete the order form, and the copyright permission will end up in your shopping cart. If you reproduce copyrighted content in a book (for example, the list starting on page 55 that somehow exceeds the notion of fair use), you must ask the Publisher of the book to purchase the reproduction rights. Note that the price for non-profit use ("academic") is substantially lower than for commercial use.

The ease of `copy/paste` text blocks into a collection of digital notes can lead to problems, especially when sentences and paragraphs are copied without the originating reference (full journal reference, or URL). Hence, authors should keep a tidy system with clear separation of own notations and copied phrases (for example, by systematically using boldface for the latter). But also oral communications (private as well as at meetings) can easily lead to "forgetting" the source of information.[84]

I return to the first case mentioned in Section 13.7, where I was struck by one (and the only one) complete paragraph without any typographical errors. Asked from where this passage was taken, the author replied that he did not know. One sentence was quite intriguing because of erroneous syntax with a needless double negation: *"We note that the choice of an incorrect annual alias value usually has no little or no effect on the subsequent search for other frequencies."* Passing this phrase through ITHENTICATE[85], the system returns a 96% similarity index with a sentence in Breger *et al.* (2005). Correcting "has no little or no effect" to "has little or no effect" yields a 100% match. **The idea or principle expressed in this sentence is very explicit, and should thus not be copied without reference to its source text**. There are two ways to correctly handle this situation: either by quoting the sentence *verbatim* as it is done here, or by rephrasing it in the sense *"As Breger et al. (2005) emphasize, the choice of an incorrect annual alias value usually has little or no effect . . . "*.

This example illustrates the maxim that even duplicators need careful proofreading, as is very well known from experience over the ages of copying by scribes. Perhaps the most famous example of a similar copying error is the so-called *Wicked Bible* published in 1631 by the royal printers in London, where for the seventh commandment the word "not" was omitted, resulting in *"Thou shalt commit adultery"*.[86] The books were consequently destroyed, and only a handful copies escaped, now making these books very expensive collector's items.

---

[83]http://www.copyright.com/

[84]Tacit knowledge and cryptomnesy are explained in Section 7.10 of Paper III.

[85]CROSSCHECK: www.crossref.org/crosscheck.html (see an example in Sect. 16.5 of Paper III).

[86]Exodus 20:14.

13.14   How to cite or quote a straying paper?

When using secondary sources, like review papers, it often happens that you cannot access the quoted primary information in that review. In such a case, you should be careful not to substitute the reviewer's opinion for the views expressed in the primary source. As an example, I refer to Section 2, where I cite Perkowitz (1989), although I have not been able to consult the original quotation – even by tracing and contacting the copyright holder, who took two months only to answer *"If you would kindly provide your Visa or MasterCard number and the expiration date..."*. Therefore, the reference at the end of the paper contains the note "quoted by Kuyper (1991)". This is quoting from second-hand information in the absence of primary sources: it is less secure, but it does not mislead the reader, and it is acceptable because it is fully documented. The same approach was followed when citing an unavailable paper, see page 35 referring to Giuseppe Baretti.

13.15   How to sort words that incorporate diacritics?

The alphabetic sort order depends on language, but also on culture – even within the same language family. Moreover, some languages use *diacritics*: marks below, above or through a character, or a combination of characters by which the sound of the letter is changed. Examples are č, ç, œ, ø, ö, ñ: these are different characters, not different fonts of a same character – see, for example, the short sentence in Czech language on page 35. The ASCII character set does not contain any diacritics, hence an ASCII sort will lead to problems, unless assisted by manual intervention. In English, a letter with diacritic is sorted after the letter without diacritic. But should we sort ae–af–æ or ae–æ–af? The *Chicago Manual of Style* (15th edition) says that for English-language sorting, words beginning with or including accented letters are alphabetized as though they were unaccented: see, for example, the position of Öpik in the Name Index at the end of this book.[87] Some letters with diacritics can be spelled without them, for example u and o with umlaut (ü, ö) can be written, respectively, as ue and oe, which is fine as long as the same spelling is applied throughout the text, and in the reference and index lists.

Sorting Greek-letter names is a special case: the LaTeX automatic sorting produces a list alphabetized according to the English spellings of the Greek letters, rather than in the order of the Greek alphabet. But when sorting star identifications in which a specific star is identified by a Greek letter, followed by the genitive form of its parent constellation's Latin name,[88] sorting should first be according to the constellation name, followed by an alphabetization following the Greek alphabet.

Sorting is also important when using acronyms, such as those of photometric systems. The Johnson *UBV* photometric system[89] has three passbands with very specific wavelengths progressing from blue to red, and the same is true for the Strömgren *uvby* system, where the central wavelengths run from 350 to 550 nm. But a GOOGLE search for *uvby* yields about 25 000 hits, while a search for the non-existing *ubvy* yields 8 000! Some

---

[87]This feature is mostly incorrectly rendered when sorting in LaTeX.

[88]The so-called Bayer designation established by Johann Bayer in 1603.

[89]These examples refer to systems designed by Harold Johnson, Bengt Strömgren and Theodore Walraven.

photometric systems list their bands in opposite sequence, for example the Walraven *VBLUW* thus follows a wavelength progression from red to blue in its acronym. Some designations have diacritic characters (subscript or superscript indices), or mix upper and lower case letters. It is most important not to change the pre-defined order or the case of these designations (see also the genealogy tree in Fig. 13 of Paper II).

13.16　To cite the URL or to cite the complete bibliometric reference?

The consequences of all too quickly pasting an URL were described in Section 11.9, and even if the URL link delivers the correct paper, the `astro-ph` reference referred to in Section 13.8 does not always lead to the latest and correct version of the accepted manuscript. So, even importing reference lists does not exempt the authors from verifying the entries. Not to speak of even more confusing statements like *"This article originally appeared in Icarus Vol. 187/1, March 2007, pp. 132–143. Those citing this article should use the original publication details."*[90]

13.17　Why publish your data?

There are several reasons for publishing or for publicly archiving primary data:

1. some papers use observational data that are defective in terms of standardization and calibration, and such flaws may lead to incorrect conclusions (see Sterken 2011 for an example). Hence, the publication of primary data may provide the basis for rectification eventually;

2. data sets – especially time series data, but also spectral energy distributions – are always incomplete, hence inclusion of data in a publication will allow subsequent studies to fill in missing phases, and even to finetune the derived frequencies;

3. archived data sets allow for meta-analysis, examples abound in Virtual Observatory projects (see, for example, Solano 2006);

4. data, especially historical archives, may later be used in ways that the originator of the dataset never could have imagined;

5. sharing detailed research data is associated with increased citation rate, see Piwowar *et al.* (2007), and also Section 18.4 in Paper III;

6. preservation of data provides the option of avoiding duplication of expensive experiments;

7. at the same time, the availability of original research data allows reconstruction of the theory of an instrument with which previous observations were collected;

8. calibration of astrophysical parameters (like effective temperature or metallicity as a function of color index) often rely on model-atmosphere calculations, and without the numerical model data, such calibrations can never be considered for upgrading;

---

[90]Footnote to the title of the paper by Belton *et al.* (2007) reproduced in Icarus Vol. 191, 2007, pp. 310–321, see also Section 7.12 in Paper III.

9. computer code should also be published in full (even though future machines will not be able to handle the code format) because a well-documented published code can be of great value in designing new data-reduction and analysis packages.

13.18   Why do journals engage in publishing data?

There is a substantial difference between a *data paper* (or data in a paper) and a *data archive*. The latter merely is a set of tabulated information, the former usually also elaborates why the data were obtained, explains how the data were acquired, includes a discussion of the accuracy, and possibly gives some illustrations of results already published elsewhere. Journals, through the organized structure of the editorial flow, and supported by verification through peer review, are gatekeepers of good scientific practice. They sustain agreements with archives (like the accord between A&A and CDS, see the papers by Agnès Henri and Laurent Cambrésy in Part 1 of these Proceedings) that assure that archived data are solidly linked with the originating journal article, that the data are rightly formatted with a correct description of the metadata,[91] and that they bear a robust time stamp – that is, that the content of the web archive is not modified *a posteriori* without proper documentation.

It is always good practice to check the journal's policy on enclosing data before you submit a paper.

13.19   How to cope with writer's block?

Writer's block or *writer's anxiety* is the inability to complete a manuscript due to many reasons. A major ground can be self doubt, fear of failure, or the fright of making significant errors. But also the seeking to produce a perfect prose at the first try, or being crushed by deadlines. A frequently occurring problem is information overload, effectively blocking the organization of thoughts and research notes. Writer's block occurs to all classes of text writers: fiction as well as non-fiction, song writers, poets, movie-script authors and journalists.

There are strategies or techniques to escape out of such dead-end streets, for example by first tracing a mind map or writing an outline, then by composing the first draft, and subsequently by reworking and polishing. A good start is by following the checklist presented on page 55. And if nothing helps, put the job aside for a while, and concentrate on something completely different. Or, look for a fine *Muse*[92] to help drawing out of you the first paragraphs of your opus.

Severe writer's block syndrome occasionally occurs when one author (usually the principal author) feels uneasy about the presence – or the rank in the authorlist – of one or more somewhat inert coauthors. The misgivings about such inactive associates can render any creative labor impossible, effectively creating a block that unavoidably will turn into a total stop – and the more pressure these inert people exert, the more massive

---

[91] See Laurent Cambrésy's paper in Part 1 of these Proceedings.
[92] In Greek mythology, the Muses are the nine goddesses who inspire artists and writers.

this block will develop. From my personal experience I know that none of the above-mentioned hints – not even a choir of Muses – could possibly solve such a mental block. You should always realize that you are not the first to live through such a situation, and you are definitely not the last one to undergo it, and that there is only one solution: just sit down and write, so that you can turn the page on that project as fast as possible, and look forward to working with other people. Or quit science.

13.20   Should I allow someone else to completely rewrite my manuscript?

Never! But there are always ways to make a manuscript more readable by rewriting some sentences or even complete paragraphs. If

1. the manuscript has a proper structure (as explained in Sect. 9);
2. the language is straightforward and direct, without winding digressions;
3. the principles explained in Adams' and Adams, Halliday, Peter & Usdin's contributions in Part 1 of these Proceedings, have been followed;
4. careful and multiple proofreading was done;
5. correct citation practice was implemented (including verification of references); and
6. you ran the text file through a good spelling checker,

then I see no reason whatsoever to have your manuscript completely rewritten by a post-doc or a supervisor.

Such forced rescripting may lead to damaging consequences. First of all, the rewriter may claim a higher rank in the list of authors. Second, if the senior rewriter was not involved in all aspects of the research, he or she may change your conclusions, which in turn leads to a modification of accountabilities. Third, the young scientist's reputation is harmed in the long run – even if only by the brunt of gossip. And, worst of all – as I experienced myself – the terrible feeling so well expressed by Melanie Safka:

> LOOK WHAT THEY DONE TO MY SONG MA ... IT IS THE ONLY THING THAT I COULD DO
> HALF RIGHT ....

PhD supervisors should manifestly avoid such scenarios. **It is the student's indisputable right to stand by their song – that is, if it is truly the product of their own creativity, and if it is at least half right in each of the 6 items listed in the previous paragraph.**

Rewriting does occur in the commercial press, for example TIME Magazine states that *"Letters may be published in Time Magazine and edited for purposes of clarity and space"*. And it happens, in a milder form, by Editors of conference proceedings, see Section 13.21.

13.21   What to do when disagreeing with editorial changes?

There are three levels of editorial changes: straight typographical errors in the body of the text (including formatting errors in the references), linguistic changes, and outright censorship.

Typographical and formatting errors, if detected, are normally corrected in the background by the Publisher's copy editor or by the language editor, *i.e.*, without asking the author's consent. Such changes – sometimes referred to as *language washing* – are almost always justified, and should not be disputed. But be careful: in the most recent book I coedited (Milone & Sterken 2011), the copy editor had replaced all mentions of CCD (Charge Coupled Device) by CD (Compact Disc?), probably because that editor's dictionary file suggested this replacement. This only illustrates that you should proofread again after receiving the page proofs.

Changes in style and syntax are more tricky, because the language editor may misunderstand what the author exactly wants to say. Moreover, the language corrector, in most of the cases, does not hold a degree in astronomy or in another science,[93] and this can lead to misconceptions. I have had only once in my career a disagreement with a language editor: in Sterken & Jaschek (1996) I described a light curve with a so-called *stillstand* phase.[94] The Cambridge University Press language corrector changed all occurrences of that term into the linguistically correct *standstill*. I then argued with the Editor that the stillstand phenomenon had been historically coined as such, and that it had become a jargon[95] word just like *bremsstrahlung* or *metallicity*[96]. The Editor accepted, and the phrasing appeared in the way I intended. For almost all other suggestions, I was quite happy with the language editing because it improved the readability of my paper, or because I could not find a better wording than the one suggested by the language editor. It is, nevertheless, important that the author has the opportunity to discuss, or even disagree with some of the proposed changes. Stronger even, such changes should never be made in the background, *i.e.*, they should be made **by** the author, and not **for** the author.

Outright censoring, without the author's consent or without the author's knowledge, should never occur. I experienced it twice in conference proceedings: once an unfavorable citation was removed, the second case even involved the removal of an entire Figure and the discussion connected to it. Since I only discovered this when the books were delivered almost one year later, protests were to no avail. Such behavior is rightly considered as editorial misbehavior, see Section 10.2 in Paper III.

So, how must such conflicts be handled by the author? The journal Editor should be the first person to contact in case of disagreement with the language corrector. Just like answering the referee, you must take care that your complaint embeds the first two storytelling elements referred to on page 2: the *content* of the message, and the *form* of the message. If the form is bad (rude language, many typos), you will never be able to convince anyone that you know better about syntax and style than the language editor does. Likewise, if the content falls short in terms of argument or proof, you cannot win over the Editor.

---

[93]Claire Halliday, one of the language editors at A&A, holds a PhD in astronomy.

[94]A plateau on the rising branch of the light curve of the pulsating star BW Vulpeculae.

[95]Paper II deals with similarly "wrong" jargon: the using of *font* for *typeface*.

[96]Metallicity refers to the proportion of matter in a star made of elements other than H and He.

13.22   Is it too much to expect authors to read the papers they cite?

This question is a *rhetorical question*[97] by Young (1993a), who pointed out that by mis-quoting, an erroneous formula had been attributed to him. A few months later, he reports to have found the same incorrect equation in another report published since. He thus con-cludes (Young 1993b) that it is evident that the latter authors have simply copied both the formula and the reference from the previously mentioned paper, without checking if either was correct. He subsequently states to have no objection when those who do not understand the physics use an incorrect formula, but that he does object to having the wrong formula attributed to him. Hence his concluding question

IS IT TOO MUCH TO EXPECT AUTHORS TO READ THE PAPERS THEY CITE?

This question also bears on Öpik's (1977) own story: a critic first completely misrepre-sents his work, and then – rightly – sets out to destroy this fantasy of his own imagination (which was just the very opposite of what Öpik was saying). So, in order to avoid the cre-ation of a myth on the basis of second-hand information, Öpik (1976) published a letter of rectification in *Science*, which was immediately followed by a rather vague and irrelevant rebuttal by Ulrich (1976).

Young's example refers to copying mistakes from journal to journal, and such cases can be easily traced. Similar copying errors from journal to book, and in book translations, do frequently occur, and are much more difficult to spot. One such case – unfortunately again with a formula by A. T. Young – was spotted by Tuvikene (2008) while using a formula for atmospheric extinction[98] copied from Young (1974) by Hall & Genet (1982). Interestingly, the latter authors give the complete and correct bibliographic reference to Young (1974), though that reference cannot be found in ADS, which only gives the *bib-code* of the book in which Young's contribution is a chapter. As Hall & Genet (1982) is not any more readily available in libraries, and since they did not quote the primary source (Young & Irvine 1967), it was not possible to find and check the original formula, and so an erroneous extinction correction that was programmed in a computer code had to wait for detection and correction during almost half a century!

Note that erroneous results in both examples would not be visible at face value, but would become appreciable in the somewhat extreme limits of the parameter domain wherein these formulae were applied.

13.23   How to review a paper?

If you are asked to referee a paper, the first thing you must ask yourself is if you are really competent in the matter. The second point is whether you can comply with the time that the Editor offers for completing the review (typically three to four weeks for a normal research paper, much less for a Letter). Finally, you must check whether there is any *conflict of interest*, *i.e.*, if you have any stake that compromises the action of being a

---

[97] A figure of speech in question format, without expecting an answer.
[98] $X = \sec z[1 - 0.0012(\sec^2 z - 1)]$ but with the exponent 2 in the argument $\sec z$ omitted.

reviewer. Obstacles could be that you are in strong competition with the authoring team, or even that you are (or have been) part of the project (see the case discussed in Sect. 8.7 of Paper III), or that the other author is a member of your institute, or is one of your best – or worst – friends. If you have the competence, if there is no conflict of interest, and if you have the time, then you may accept the invitation.

I have nowhere seen explicit guidelines, but a most helpful guide for critical assessment of a research article is the *"checklist for critiquing a research article"* by Kuyper (1991): she gives a structured list – following the IMRAD scheme – of 20 cardinal points to be considered when reviewing a paper:[99]

- INTRODUCTION

    1. Read the statement of purpose at the end of the introduction. What was the objective of the study?

    2. Consider the title. Does it precisely state the subject of the paper?

    3. Read the statement of purpose in the abstract. Does it match that in the introduction?

    4. Check the sequence of statements in the introduction. Does all information lead directly to the purpose of the study?

- METHODS

    5. Review all methods in relation to the objective of the study. Are the methods valid for studying this problem?

    6. Check the methods for essential information. Could the study be duplicated from the information given?

    7. Review the methods for possible fatal flaws. Is the sample selection adequate? Is the experimental design appropriate?

    8. Check the sequence of statements in the methods. Does all information belong in the methods? Can the methods be subdivided for greater clarity?

- RESULTS

    9. Scrutinize the data, as presented in tables and illustrations. Does the title or legend accurately describe content? Are column headings and labels accurate? Are the data organized for ready comparison and interpretation?

    10. Review the results as presented in the text while referring to data in the tables and illustrations. Does the text complement, and not simply repeat, data? Are there discrepancies in results between text and tables?

    11. Check all calculations and presentation of data.

---

[99]Reproduced with permission after purchase by Publisher EDP of reproduction rights from the copyrights holder.

12. Review the results in the light of the stated objective. Does the study reveal what the researcher intended?

- DISCUSSION

13. Check the interpretation against the results. Does the discussion merely repeat the results? Does the interpretation arise logically from the data, or is it too far-fetched? Have shortcomings of the research been addressed?

14. Compare the interpretation to related studies cited in the article. Is the interpretation at odds or in line with other researchers' thinking?

15. Consider the published research on this topic. Have all key studies been considered?

16. Reflect on directions for future research. Has the author suggested further work?

- OVERVIEW

17. Consider the journal for which the article is intended. Are the topic and format appropriate for that journal?

18. Reread the abstract. Does it accurately summarize the article?

19. Check the structure of the article (first headings and then paragraphing). Is all material organized under the appropriate heading? Are sections subdivided logically into subsections or paragraphs?

20. Reflect on the author's thinking and writing style. Does the author present this research logically and clearly?

Not only can most of these 20 points be directly applied to astronomy, the list is also an excellent guide or blueprint in planning the writing of a paper.

13.24   How to revise a multi-authored paper?

This is a matter of optimizing discussion and communication, a prime condition for successful collaborative writing. For a paper with multiple authors, there are three ways to deal with the referee report, and the feedback on it:

1. THE HUB-AND-SPOKE NETWORK: a system where all communication traffic moves along spokes connected to the central hub, which is the corresponding author;

2. THE MERRY-GO-ROUND METHOD: sending the manuscript source file down the list of authors, so that every contributor makes changes and then sends the file to the next coauthor;

3. THE PARALLEL COMMUNICATION MODEL: the first author sends the referee report as a group mail to all coauthors, who (normally) return their feedback in parallel lines with copy to everyone involved.

The first method has the disadvantage that it is slow, that information is only partially shared, that the hub (you) is the bottleneck in the network, and that a short inactivity or failure of the hub will halt the complete system – just like a router failure in a communication network between computers will stop all communication. But the major drawback is that information exchange is not open, that there is no archiving, and that *clique* forming[100] is more likely to occur. Once such a situation is established, it becomes much more difficult to formulate a coherent response towards opposing views of a referee.

The second *carousel* course is the least creative, and results in papers of the lowest finishing quality. Moreover, it is utterly inefficient, and not seldomly leads to multiple people simultaneously modifying the same source files. Not to mention that not every one uses the same text editor or LaTeX settings, nor do we all have the same level of skill in typesetting and managing LaTeX.

The third approach, on the other hand, is more frank, since each coauthor can take note of everyone else's opinions. It also provides a straightforward view of who is really active in the team.

I recommend by far the parallel approach, and I have seen many a revised paper lagging because the *hub* inadequately filtered (*i.e.*, read, misread, translated or transferred) the opinions of one author to the other. The parallel route cuts down on possible disagreements between subsets of coauthors. The parallel communication model provides a strong edge in handling the problems referred to in Section 13.21, notably when dealing with language editing. Imagine, for example, that there is only one native English-speaking person among the half a dozen coauthors of a multi-authored paper, and that all editorial changes are taken care of by the first author whose mother language is not English. Even though language editing, as a rule, does ameliorate syntax issues, it may happen that slight changes in particular idiomatic expressions (such as *stillstand*, see page 53) are suggested, and that such nuances are not detected by the corresponding author, while they would be noticed by the native English-speaking coauthor if the parallel communication model would have been implied. Thus, without feedback from all authors – even at the language-editing level – the published paper may in the end contain unwanted expressions, and thus arouse feelings of frustration in all parties involved.

Figure 2 of Paper II illustrates noise sources in the model of information transmission, and each of the three communication scenarios described above contributes its own level of noise on the sender's side of the communication train. It is of utmost importance to keep all those noise levels as low as possible.

13.25   When can I ask for another referee?

Asking for another referee is a right, however, you must first make sure that your request is based on solid grounds, *e.g.*, if you think that the referee is plainly wrong, or if the report contains rude personal attacks. And really think twice before you act. For example, in the first case mentioned in Section 13.7, the first author suggested to ask for a new referee, a

---

[100] A subset of the coauthor population may form a common view of the other – or even just one – coauthor(s).

request in disagreement with the second author's opinion, resulting in withdrawal of the manuscript.

In general, it is not wise to start a head-on fight with a referee, unless you are sure that you are right and the reviewer is darn wrong: it is better to display some diplomatic modesty in the way Wright & Armstrong (2007) advise:

> ... IF I BELIEVE A REFEREE IS MISTAKEN IN HIS/HER CONCERN, AND I KNOW A WAY TO DEFUSE THAT MISTAKEN CONCERN WITHOUT TELLING THE REFEREE THAT HE/SHE IS MISTAKEN, THEN I WILL USE THAT WAY BECAUSE THE PROBABILITY OF SURVIVING THE REVIEW PROCESS DECREASES WHEN REFEREE CONCERNS ARE CHALLENGED RATHER THAN ACCEPTED.

But if you decide to oppose to the referee's views, remember from Section 13.9 to formulate your arguments in a non-offensive way, otherwise you will be judged on the form of the message, and not on its content, and you will lose the battle. As Wright & Armstrong (2007) rightly point out, it is a matter of surviving the revision process.

13.26   Why do Editors edit, and why do referees review?

Editors of non-commercial journals do not operate on a classic *pay-for-work* scheme, though they receive indemnities for the editorial work that they carry out besides their regular research or teaching functions. Nor do referees receive any remuneration. So why do they bother to do it? Well, Editors and referees receive very interesting papers to review in the first place, and this leads to professional enrichment, and helps them keeping up with the advancements in their fields.

I have once witnessed a senior referee presenting his "bill" at a meeting of Editors and Publishers: taking one work day for one paper to review, with about a dozen papers per year to referee, results at the end of a career in about two years of *work-without-pay*. Although I concur with the calculation of labor time,[101] my personal vision is that I still consider myself to benefit a major profit in terms of new facts learned from these in-depth readings. Moreover, I consider refereeing a form of service to the community, and even an effective mode of parental investment towards the students. **Contributing to the editorial process through refereeing should never be the subject of a cost-benefit analysis**, and I am convinced that when money becomes a prime motive for refereeing, the system will then really break down. Editors need independent referees because, without reviewers it is simply not possible to run a high-quality journal, see also Section 8 (on referee misconduct) in Paper III.

Reporting a new scientific result always implies imaginative thoughts (hypothesis, speculation, ...) and criticism (error budget, validation ...). It is very often difficult to produce yourself an exhaustive and complete critical analysis of all aspects of your own scientific reasoning. So, criticism or judgment by an external person (or by a competitor) – just like proofreading by alien eyes – may bring out deficiencies that escaped the authors' attention (or that were poorly explained). The external referee is just part of the

---

[101] Klahr (1985) estimates the total cost of one panel review to be equal to 3 person-days.

cooperative and open enterprise called science, and merely adds to the dialogue between imagination and critical judgment, with the referee's feedback not seldomly constituting a true bonus to the information content and credibility of the reviewed paper.

14 Summary

Communication of science is as important as the research results themselves. Top results just get higher exposure when they are transmitted in a *state of the art* fashion. Sandage (1989) underlined this most properly for Edwin Hubble's writing style:

> IT WAS HUBBLE'S MASTERY OF THE LANGUAGE THAT GAVE SOME OF HIS PAPERS SUCH DOMINANCE OVER PRIOR WORK BY OTHERS. OFTEN THE PROBLEM HAD IN FACT BEEN SOLVED, BUT WITHOUT THE SAME ELEGANCE OF STYLE, POWER OF PRESENTATION, AND EXCELLENCE OF SUMMARY POSSESSED BY HUBBLE WHEN HE WAS AT HIS BEST. CLEARLY, THE LESSON FOR STUDENTS IS LEARN TO WRITE AT THE SAME TIME THAT YOU LEARN TO DO GREAT SCIENCE.

But Hubble was not just anyone: he had received a formal training as a lawyer, and as such had learned how to clearly explain evidence, hypothesis and conclusion to his opponents. None of us can possibly dream of having such a background, but we can take a lesson from lawyers: **know your facts, understand your data, explain your points, and learn from your adversaries**. And, if you receive comments on your manuscripts, then

1. learn to properly handle invaluable feedback, and try not to get upset or feel offended: many remarks are fair, and most are made to help you;

2. understand that errors are normal for beginners (even for old hands), that failure is a necessary aspect of research, and that the best way to learn is to learn from your own and from other people's errors;

3. know that most Editors and referees try to help you, and that most papers get published after all.

Above all, exercise self-examination, with respect to the scientific content and significance of your paper, as well with regard to the presence and rank of your name in the author list. And never forget that it is quite normal that writing an innovative paper always takes much more time than scribbling a copy-and-paste article.

Postscript

In Sterken (2006) I explain that the comma serves to indicate a pause, and that omission or transposition of a comma may completely change the meaning and the direction of a phrase. I demonstrated the alteration of the meaning by using the following sentences in which only the number and place of the commas differ:

> CHRIS SAID, CONNY LET'S MEET TOMORROW, and
> CHRIS, SAID CONNY, LET'S MEET TOMORROW.

I used the same example in a lecture in Paris,[102] and after my talk a colleague came to me, and said

> YOUR TALK SADDENS ME: IF STUDENTS NEED TO LEARN EVERYTHING ABOUT COMMAS
> AND PUNCTUATION MARKS, WHERE WILL THEY EVER FIND THE TIME TO DO SCIENCE?

I answered that the education in scientific writing is like the grooming in *haute cuisine*, where excellence is not only achieved in elaborate preparations of exquisite food, but also in avoiding artificiality and pretension when the food is served.

Molière expressed this aspect of punctuation and syntax so well in his comédie-ballet *Le Bourgeois Gentilhomme*,[103] in which Monsieur Jourdain asks his teacher for help rendering the sentence *"Beautiful marquise,[104] your beautiful eyes make me die from love"* in a more elegant way. The *Maître de Philosophie* then offers him five sentences that differ only in the emplacement of the words, and in the punctuation marks, see Figure 4. Each of these sentences has exactly the same meaning – that is, as long as the punctuation marks are correctly placed. In the end, Monsieur Jourdain asks his teacher which formulation is the most elegant, and the reply simply was: **the very first one you proposed**.

That story underlines that you are not expected to write like Shakespeare, nor like Molière, but that your writings should honor the basic characteristic of scientific style as explained on page 12: your style should be direct, plain, assertive, context-independent, and brief.

Belle Marquise, vos beaux yeux me font mourir d'amour.
↑

D'amour mourir me font, belle Marquise, vos beaux yeux.
↑ ↑

Vos yeux beaux d'amour me font, belle Marquise, mourir.
↑ ↑

Mourir vos beaux yeux, belle Marquise, d'amour me font.
↑ ↑

Me font vos yeux beaux mourir, belle Marquise, d'amour.
↑ ↑

Fig. 4. Five phrases with the same words rearranged, though with different punctuation as indicated by vertical arrow symbols. Source: Molière, *Le Bourgeois Gentilhomme* (1670). Typeface Arial, font Arial 18 pt, reduction factor 50% (see Paper II).

I am grateful to PhD students Wu Yue and Zhang Li Yun for kindly providing the Pinyin transliterations reproduced in Figure 2, and to Veerle J. Sterken for proofreading a first version of this manuscript. I thank the Publisher EDP for purchasing the reproduction right to list the 20 points in Section 13.23.

---

[102]*Comité pour les Pays en Développement*, Institut de France, Académie des Sciences, 3 December 2009.
[103]*The Bourgeois Gentleman*, by Jean-Baptiste Poquelin (1622–1673).
[104]A *marquise*, or marchioness, is a noblewoman ranking below a duchess and above a countess. But this word has also a second meaning: a *marquise diamond cut* is one of the more elegant (oval) diamond shapes.

References

Allen, R.J., 1970, Stephan's quartet?, A&A, 7, 330

Belton, M.J.S., Thomas, P., Veverka, J., *et al.*, 2007, The internal structure of Jupiter family cometary nuclei from Deep Impact observations: The "talps" or "layered pile" model, Icarus, 187, 332

Berthold, G., 1897, Ueber den angeblichen Ausspruch Galilei's, Bibl. Math. p. 57

Bertout, C., & Schneider, P., 2004, Introducing structured abstracts for A&A articles, A&A, 420, E1

Bertout, C., & Schneider, P., 2005, Editorship and peer-review at A&A, A&A, 441, E3

Bottle, R.T., 1973, Information Obtainable from Analyses of Scientific Bibliographies, Library Trends, 22, 60

Breger, M., Lenz, P., Antoci, V., *et al.*, 2005, Detection of 75+ pulsation frequencies in the d Scuti star FG Virginis, A&A, 435, 955

Dobell, C., 1939, Dr. O. Uplavici (1887–1938), Isis, 30, 268

Drenth, J.P.H., 1998, Multiple Authorship: The Contribution of Senior Authors, J. Amer. Med. Assoc., 280, 219

Dubois, P., 1995, personal communication, 16 November 1995

Duerbeck, H.W., 2005, Book Review: Astronomy in and around Prague, J. Astron. Data, 11, 5

Feynman, R.P., 1985, Surely You're Joking, Mr. Feynman! (New York: W.W. Norton & Co)

Hall, D.S., & Genet, R.M., 1982, Photoelectric Photometry of Variable Stars (Fairborn, Ohio), p. 9-4

Hames, I.., 2011, EASE forum, http://www.ease.org.uk/forum/index.shtml

Klahr, D., 1985, Amer. Psychol., 40, 148

Kozak, M., 2009, Eur. Sci. Edit., 35, 102

Kuyper, B.J., 1991, BioScience, 41, 248

Lauer, T.R., Faber, S.M., Ajhar, E.A., *et al.*, 1998, AJ, 116, 2263

Lemaître, G., 1927, Un Univers homogène de masse constante et de rayon croissant, rendant compte de la vitesse radiale des nébuleuses extra-galactiques, Annales de la Société scientifique de Bruxelles, XLVII, 49

Lepschy, L., 2006, personal communication

Li, C., Fu, J.-N., Zhang, Y.-P. & Jiang, S.-Y., 2007, A new study of the long-term variability of the SX Phoenicis star CY Aquarii, J. Beijing Normal Univ., 43, 621

Malthus, T., 1798, An Essay on the Principle of Population, London

Matzinger, P., & Mirkwood, G., 1978, J. Exper. Med., 148, 84

Maunder, A.S.D., 1902, MNRAS, 62, 57

Maunder, W., Mrs., 1904, MNRAS, 64, 224

Maunder, A.S.D., & Maunder, E.W., 1908, The heavens and their story (Robert Culley, London)

Merton, R.K., 1995, Social Forces, 74, 379

Milone, E. F. & Sterken, C. (eds.), 2011, Astronomical Photometry: Past, Present, and Future (New York: Springer Verlag)

Oort, J.H., 1970, A&A, 7, 405

Öpik, E.J., 1976, Solar Models, Science, 191, 1292

Öpik, E.J., 1977, About Dogma in Science, and Other Recollections of an Astronomer, Ann. Rev. Astron. Astrophys., 15, 1

Perkowitz, S., 1989, Commentary: can scientists learn to write?, J. Techn. Writing Comm., 19, 353 (quoted by Kuyper 1991)

Pinker, S., 1994, The Language Instinct (Penguin Books)

Piwowar, H.A., Day, R.S., & Fridsma, D.B., 2007, PLoS ONE www.plosone.org, doi: 10.1371/journal.pone.0000308

Pottasch S.R., & Steinberg, J.L., 1969, A&A, 3, 381

Qiu, J., 2008, Identity crisis, Nature, 451, 766

Salager-Meyer, F., 2009, Academic equality and cooperative justice, TESOL Quarterly, 43, 703–709

Sandage, A., 1989, Edwin Hubble 1889–1953, JRASC, 83, 351

Shermer, M., 2007, Scientific American, October 2007, 23–24

Sheskin T.J., 2006, Sci. Eng. Ethics, 12, 555

Schielicke, R.E. (ed.), 1998, In Reviews in Modern Astronomy 11: Stars and Galaxies (Hamburg, Germany: Astronomische Gesellschaft), 1

Shore, S.N., 2009, A&A, 500, 31

Sokal, A., 1996, Transgressing the Boundaries: Towards a Transformative Hermeneutics of Quantum Gravity, Social Text (Duke University Press), 46/47, 217

Solano, E., 2006, Virtual Observatory: What Is It and How Can It Help Me?, in Astrophysics of Variable Stars, ed. Sterken, C. and Aerts, C. (San Francisco: Astronomical Society of the Pacific), ASP Conf. Ser., 349, 29

Stein, K., 2006, J. Amer. Diet. Assoc., 106, 1947

Sterken, C., 1977, A&A, 57, 361

Sterken, C., 1988, The Coordination of Astronomical Observations in Coordination of Observational Projects in Astronomy, ed. C. Jaschek & C. Sterken (Cambridge University Press), p. 3

Sterken, C., 2000, Mode Identification through Photometry: Full Colours?, in Delta Scuti and Related Stars, Reference Handbook and Proceedings of the 6th Vienna Workshop in Astrophysics, ed. M. Breger and M. Montgomery (San Francisco: ASP), ASP Conf. Ser., 210, 99

Sterken, C., 2005, The Light-Time Effect in Astrophysics: Causes and Cures of the $O - C$ Diagram, ed. Sterken, C. (San Francisco: Astronomical Society of the Pacific), ASP Conf. Ser., 335, 178

Sterken, C., 2006, Advice on Writing a Scientific Paper, in Astrophysics of Variable Stars, ed. Sterken, C. and Aerts, C. (San Francisco: Astronomical Society of the Pacific), ASP Conf. Ser., 349, 445, http://www.aspbooks.org/publications/349-0445.pdf

Sterken, C., 2007, On the Provability of Heliocentrism. I. Ole Roemer and the Finite Speed of Light, Sartoniana, 20, 19

Sterken, C., 2011, Just one new measurement of the B[e] supergiant Hen-S22, IBVS, 6000

Sterken, C., & Aerts, C. (eds.), 2006, Astrophysics of Variable Stars, ASP Conference Series, Vol. 349

Sterken, C., & Manfroid, J., 1992, Astronomical Photometry: a Guide (Kluwer Academic, Dordrecht)

Sterken, C., & Jaschek, C., 1996, Light Curves of Variable Stars, A Pictorial Atlas (Cambridge: Cambridge University Press)

Sterken, C., Samus, N., & Szabados, L. (eds.), 2010, Variable Stars, the Galactic Halo and Galaxy Formation (Moscow: Sternberg Astronomical Institute of Moscow University), p. 125

Sweetland, J.H., 1989, Errors in Bibliographic Citations: A Continuing Problem, The Library Quarterly, 59, 291

Thomas, K., 1991, in A Cultural History of Gesture: from Antiquity to the Present Day, ed. J. Bremmer & H. Roodenburg (Polity Press, Cambridge), p. 1

Tukey, J.W., 1977, Exploratory Data Analysis (Addison-Wesley Publishing Company)

Tuvikene, T., 2008, personal communication

Ulrich, R.K., 1976, Science, 191, 1293

Watson, J.D., 2007, Avoid Boring People (Oxford University Press)

Wright, M., & Armstrong, J.S., 2007, Verification of Citations: Fawlty Towers of Knowledge?, Munich Personal RePEc Archive, 4149, http://mpra.ub.uni-muenchen.de/

Wróblewski, A., 1985, de Mora Luminis: A spectacle in two acts with a prologue and an epilogue, Am. J. Phys., 53, 620

Young, A.T., 1974, Observatyional technique and data reduction, in Methods of Experimental Physics, Vol. 12, Astrophysics, Part A: Optical and Infrared. ed. Carleton, N.P. (New York, Academic Press), p. 152

Young, A.T., 1993a, Scintillation noise in CCD photometry, Observatory, 113, 41

Young, A.T., 1993b, Propagation of errors, Observatory, 113, 266

Young, A.T., 1994, in The Impact of Long-term Monitoring on Variable Star Research: Astrophysics, Instrumentation, Data Handling, Archiving, ed. C. Sterken & M. de Groot, NATO Advanced Science Institutes (ASI) Series C: Mathematical and Physical Sciences (Dordrecht: Kluwer)

Young, A.T., & Irvine, W.M., 1967, Multicolor photoelectric photometry of the brighter planets. I. Program and Procedure, AJ, 72, 945

Some additional readings:

Anderson, L.K., 2006, McGraw-Hill's Proofreading Handbook (London: McGraw-Hill)

Butcher, J., 2004, Copy-editing: The Cambridge Handbook for Editors, Copy-editors and Proof-readers third edition (Cambridge: Cambridge University Press)

The Chicago Manual of Style 2003, The Chicago Manual of Style: The Essential Guide for Writers, Editors, and Publishers, 15th edition (Chicago: The University of Chicago Press)

Sullivan, K.D., & Eggleston, M., 2006, The McGraw-Hill Desk Reference for Editors, Writers, and Proofreaders (London: McGraw-Hill)

Scientific Writing for Young Astronomers
C. Sterken (ed)
EAS Publications Series, **50** *(2011) 65–170*
www.eas-journal.org

WRITING A SCIENTIFIC PAPER II.

COMMUNICATION BY GRAPHICS

C. Sterken [1]

Abstract. This paper discusses facets of visual communication by way of images, graphs, diagrams and tabular material. Design types and elements of graphical images are presented, along with advice on how to create graphs, and on how to read graphical illustrations. This is done in astronomical context, using case studies and historical examples of good and bad graphics.

Design types of graphs (scatter and vector plots, histograms, pie charts, ternary diagrams and three-dimensional surface graphs) are explicated, as well as the major components of graphical images (axes, legends, textual parts, etc.).

The basic features of computer graphics (image resolution, vector images, bitmaps, graphical file formats and file conversions) are explained, as well as concepts of color models and of color spaces (with emphasis on aspects of readability of color graphics by viewers suffering from color-vision deficiencies).

Special attention is given to the verity of graphical content, and to misrepresentations and errors in graphics and associated basic statistics. Dangers of dot joining and curve fitting are discussed, with emphasis on the perception of linearity, the issue of nonsense correlations, and the handling of outliers. Finally, the distinction between data, fits and models is illustrated.

1 Introduction

This paper discusses elements of visual communication by way of images, graphs, diagrams and tabular material. This chapter is not intended as a manual, but rather as an elaborated field guide to scientific graphics, offering case studies and historical examples of good and bad graphics. As in Paper I, some redundancy and overlap of topics was built in on purpose, so the reader is not forced to study this paper by reading all Sections in sequence, but can consult any Section and follow the cross-references to previous and following parts in this paper and in the two associated papers.

---

[1] Vrije Universiteit Brussel, Pleinlaan 2, 1050 Brussels, Belgium

The scope of visual communication with graphics embraces all aspects of graphic design: the creative side of making graphs, the technical aspects related to the use of computer-controlled input and output devices, and the interpretive aspects when reading graphs made by other persons.[1] Section 2 explains the process of visual communication in relation to textual communication. Then follows an example of a communication without graphical content, and an example of graphics unaccompanied by text. This is followed by a detailed description of design types, and of the structural components of graphs. Subsequent Sections deal with basic aspects of graphical resolution, and with some technical features of color printing. Finally, we deal with verity of graphical content.

This paper contains several graphs copied from books and journals, and some of these graphs are reproduced on a smaller scale because they are only meant to illustrate a principle. Readers who wish to look at the scientific content of these graphs are referred to the original publication given in the list of references. This paper, as well as Paper III, also includes a selection of purpose-made graphics that accentuate the design principles explained in the text.

Why do we need such an extensive chapter on making graphs? There are several reasons.

1. An inspection of contemporary scientific graphics reveals an abundant amount of blunders: some scientific papers have even been rejected because of bad graphics (see item 11 in Sect. 5.3 of Paper III).

2. Just as authors nowadays became typesetters, they also turned into draftspersons – though with the assistance of computer graphics software. When I started my scientific career as a young postdoc, most of our graphics were made by a professional or semi-professional draftsperson who was skilled in document design, and to whom we would give a more or less tidy draft on graph paper. That person, using pen and ink, would then turn the draft design into a final graph at a scale fit for photographic reproduction. The first consequence of this division of labor was homogeneity: the person making the final product used more or less standard parameters (symbol size, line thickness, label fonts, ...). In addition, that person would tell you if you had forgotten to add tick marks, or units, or labels, and also would give you guidelines for improving your next graph. This era has definitely come to an end: young scientists are now entirely on their own.

3. A wide selection of commercial software offers integrated facilities for producing statistics from spreadsheets, and doing regression analysis and graphing in semi-automatic ways – nevertheless, many such graphs need subsequent optimization.

4. You want to persuade the reader with the best visual arguments. So, while keeping the presentations truthful and guileless, you also wish to present your product in the most pleasing and aesthetic format.

5. Although numerical quantities (and their statistical errors) from computer-aided analysis nowadays prevail, many conclusions remain based on graphical forms of

---

[1]Interpretation refers to the construction, by the reader, of the meaning of a message. That reading often differs from the designer's or the writer's intended meaning.

data presentation. Such as the position of a set of stars in some diagram: for example, the many varieties of Hertzsprung–Russell diagrams (see also Sect. 11.8) from which spectral types and evolutionary ages are derived.

6. Astrophysicists rely on calibrations and models, such as the period-luminosity relation of Cepheids, or color index versus effective temperature calibrations, and such relationships can be easily applied by means of a given formula. The example on page 162 illustrates this for spectral classification through photometry, a technique that relies on the position of a star in some diagram. The discussion shows how the graphical representation of the data tells you about the relevance or about the ambiguity of the conclusions. If conclusions are directly derived from graphical representations, it is of utmost importance to make sure that the graph correctly represents **all** the data.

7. Last but not least, just like the writing process contributes to the analysis and to the conclusions, the graphical design process sometimes reveals something that you had not yet seen in the data, equations and tabular material – some kind of *advancing insight*, as explained in Section 2 of Paper I.

2   Data graphics and visual communication

2.1   Written communication

The process of composing a scientific paper or a book often is a graphical procedure, because most scientific works consist of a body of textual elements, with embedded graphics. Paper I lists some crucial differences between oral and literate modes of communication. Communication by way of visual information, however, differs in many ways from oral and written modes of information transfer.

The mathematician Claude Elwood Shannon (1916–2001) founded information theory with his paper *A Mathematical Theory of Communication* (Shannon 1948), in which he derived the logarithmic formula for information content of a message. This technically complicated paper was followed by a much more readable book, with almost the same title[2], by Shannon & Weaver (1949). Shannon defines communication-related quantities that are "*... measures of information, choice and uncertainty*" – much akin to entropy[3] as described in thermodynamics. His simple model of a communication system is shown in Figure 1, and I use this widely copied picture as a metaphor:[4] the *information source* conceives a message, the *transmitter* transforms the message to a *signal* that is sent over a *channel* to the *receiver*, which subsequently passes it on to the *destination*. The processing of information (mainly through language) is essentially stochastic[5] in nature, and is complicated by a vexing perturbation called *noise*.

---

[2]*The Mathematical Theory of Communication*, *i.e.*, only the indefinite article was changed.

[3]Entropy is a measure of how disorganized a system is.

[4]A metaphor is a figure of speech in which a word or phrase literally denoting one kind of object or idea is used in place of another to suggest an analogy between them (from the *Merriam–Webster Online Dictionary*).

[5]A stochastic process is a process whose subsequent state is determined by a random element.

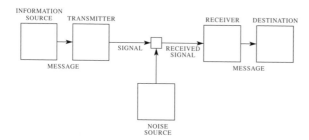

Fig. 1. Schematic diagram of a general information transmission system. Source: Shannon (1948).

In written communication, the information source and the destination are, respectively, the writer's brain and the reader's mind. In oral communication, the transmitter and the receiver are the voice and the ear. In scientific writing, these functions are taken care of by the medium that carries the message (paper or screen), and by the sensing organ (visual or touch sensation). But, whereas oral communication is person to person, scientific writing mostly emanates from multi-author composite brains. Unfortunately, the transmitting medium does not carry the original idea, but a message transformed – across time and culture – by the faculty of language, and by the inherent limitations of graphical information. Likewise, the receiving medium imprints a transformed message onto the receiver. Noise sources interfere at both ends of the chain, and contain all forms of pollution: typographical mistakes, syntax errors, translation flaws, graphics errors, chart junk, miscommunication between coauthors, etc. And the transmission channel – the journal – unavoidably adds its very own noise components: via the Editor and the referees, through language editing, via the Publisher, and through the printing process, as illustrated in Figure 2. There is, though, a fundamental difference between Shannon's communication and our transfer of knowledge: Shannon reduces communication to pure transmission of information. In scientific writing, we also deal with the **meaning** of the information, a meaning that is not only *extracted* from the transmitted message, but is subsequently also *reconstructed* by the receiver. In addition, there are several instances of **feedback**, *viz.*, from the coauthors, the Editor, and the referees.

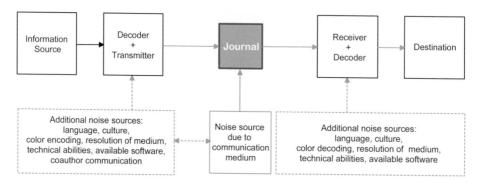

Fig. 2. Schematic diagram of an "extended Shannon" transmission model of communication.

2.2 Processing visual information

The foremost characteristic of the mode of written communication is its outspoken sequential and discrete character: words and sentences are sent and received as discrete elements in a kind of time series. The way how visual information is received and processed differs in yet another manner: the receiver (the eye) gets the message in one blink, and transfers that image to the destination (the brain) for analysis and interpretation. That assimilation of a graphical message by the brain, however, is not instantaneous, but embraces a complicated mental process that consists of breaking down the graphics in different elements. This process, in turn, leads to another time sequence of message elements – a series that differs from person to person – and that, to a considerable degree, depends on the ambience of the moment. So, just as the writing and reading processes are linked, as described in Paper I, the assimilation of graphics also engenders crucial elements of "reading" and "writing".

The opening sentence of Young's (1845) "lecture 38" on vision says:[6]

THE MEDIUM OF COMMUNICATION, BY WHICH WE BECOME ACQUAINTED WITH ALL THE
OBJECTS THAT WE HAVE BEEN LATELY CONSIDERING, IS THE EYE; . . .

Yet, a most interesting aspect of the eye as a medium of communication is that the viewer of graphical information is continuously caught in a trap of illusions caused by the very process of vision. At the same time, it is exactly through the untrue mental representation by the receiver's mind that grayscale and color printing with discrete dots is possible at all.[7]

Even more so than when producing text, the creation of graphics demands careful attention by the first element in Figure 2: the information source, or the mind. The other components in the communication chain – the transmitters (camera, draftsperson, artist, plot program), receivers and sources of noise (contrast, resolution, color, ink, . . .) – can be controlled to some extent. As photographer Patrick Demarchelier says:[8] *"Your eye is the first camera. The camera is just an accessory between you and the subject"*. In other words: **the design phase of your graphical presentation is the very first phase in the whole process**. The next step is the transfer of your technical and artistic ideas into your artwork[9]. Tufte (1983) states:

ALLOWING ARTIST-ILLUSTRATORS TO CONTROL THE DESIGN AND CONTENT OF STATISTI-
CAL GRAPHICS IS ALMOST LIKE ALLOWING TYPOGRAPHERS TO CONTROL THE CONTENT,
STYLE, AND EDITING OF PROSE.

I cannot underline enough the utmost importance of this first step: its meaning will become fully clear in Section 11, notably by the quote of Ronald Fisher (1925) on page 148.

---

[6]Thomas Young (1773–1829).
[7]Notably by continuous assimilation of high-resolution information, see Section 6.1.
[8]Time Magazine November 2008.
[9]Artwork is the general term used for all kinds of graphics.

2.3   Why use graphs at all?

Graphs and graphical illustrations are made for several reasons, *viz.*:

1. to provide a first contact with the data;

2. to facilitate communication (see, for example, Fig. 21 that visually explains different time standards);

3. to save space;

4. to detect trends and patterns (trends and geometric forms demand less visual memory than tabular information), and

5. to assist you in making decisions and interpretations.

Above all, graphical presentations should help the reader understand the facts and arguments presented in the paper.

2.4   Images, graphs and tabular structures

Artwork is just much more than a collection of pictures: **graphics comprises any illustrative material received by Publisher and Printer**, thus also non-typeset characters such as mathematical equations in graphical format, foreign language characters (*e.g.*, the Pinyin characters shown in Fig. 2 of Paper I), movies, animations, and data Tables.

Graphics come in one of three types: images, graphs and tabular structures. An image is a picture or photograph, a graph (often referred to as a plot) presents interesting data and complex ideas with precision and efficiency, and a Table gives an overview expressed by a condensed and orderly structure, with elements formatted and sorted, and very often of machine-readable conception. Technically spoken, artwork comes in two flavors:

1. *linework*: graphs of all kinds, such as diagrams, drawings, cartoons, and

2. *halftones*: "photographs" rendered in print. Movies and animations simply are time-series compositions of above-mentioned types.

3   Text without graphics and graphics without text

There is one important step that precedes Demarchelier's "first step": the decision whether to support your argument with graphics foregoes the phase of conception and design. This is an important decision, because omitting convincing graphical arguments may very well result in a paper that totally escapes attention. To demonstrate that graphics and text are intimately related, I now proceed with an example of a paper without any supporting graphics, and then I present an illustration of the opposite situation: interpreting an unfamiliar image in the absence of any textual support.

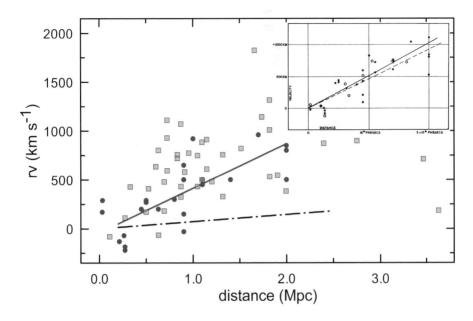

Fig. 3. Relation between observed distance and radial velocity of Hubble's "nebulae" (•: data from Hubble 1929; ■: Lemaître's available data, compiled by Duerbeck & Seitter 2001). The full line is the linear fit to Hubble's data, the dash-dotted line indicates the modern relation. The inset shows Hubble's original diagram (the dashed line in Hubble's drawing is just another way Hubble fitted his data). Source: Sterken (2009).

3.1 Consequences of non-graphical communication: an example

A painful case of non-graphical communication is the paper by the Belgian Georges Lemaître (1894–1966), one of the first relativistic cosmologists. Edwin Hubble (1889–1953) published in 1929 a list of distances[10] and radial velocities[11] of 42 "nebulae" (galaxies), and so derived the empirical velocity–distance relation as shown by the inset in Figure 3. The constant of proportionality, $H_0 = 465 \pm 50$ km s^{-1}Mpc^{-1}, was later called the *Hubble constant*.[12] Hubble proposed two solutions: one using the 42 points, and a second one based on the same data binned in groups (therefore there are two linear fits in his graph). Lemaître's theoretical deductions, based on Einstein's theory of General Relativity, and combined with the available data – in fact the same data used by Hubble – led to his vision of an expanding universe with a *linear* relation between recession velocity and redshift, and an expansion constant of 625 km s^{-1}Mpc^{-1}. Lemaître was the first who referred to a "beginning" universe.

As already explained in Paper I, Lemaître (1927) formulates his hypothesis most explicitly already in the title of his work: *Un Univers homogène de masse constante et de*

---

[10]Distances derived from the period-luminosity relation of Cepheids.

[11]Radial velocities from Strömberg (1925).

[12]The subscript in H_0 refers to the value of the Hubble constant at this moment.

rayon croissant rendant compte de la vitesse radiale des nébuleuses extragalactiques, in which he concludes that the radial-velocity data lead to an apparent Doppler effect that is caused by the change in the radius of the universe.[13] Lemaître's publication, unfortunately, appeared in the *Annales de la Société Scientifique de Bruxelles*, and thus the paper was not widely read. It is though remarkable that four years later a translation appeared "communicated by A. S. Eddington" (Eddington 1931), and that this translation does not even mention Lemaître's 625 km s^{-1}Mpc^{-1}numerical result. It thus seems that **omitting graphical representation of crucial new findings in a paper reduces its visibility and diffusion**, especially when published in a not so widely available journal.

3.2   Non-textual communication: an example

The front cover of this book reproduces, with permission of the *Röhsska Museum of Fashion, Design and Decorative Arts*,[14] the Tapestry *Astronomie* depicting the Muse *Astronomia* together with one or two astronomers, and several shepherds gazing at the heavens. It is a Flemish tapestry, probably made in Paris,[15] with linen, wool and silk. Its dimensions are 240 × 340 cm, and it was photographed by Swedish photographer Håkan Berg, who included a *Kodak Control Color Patch* strip and a *Kodak Gray Scale* calibration exposure (reproduced on the back cover of this book). A 600 dpi[16] resolution scan of an 18 × 24-cm photoprint was provided by the Museum.

Before discussing some technical aspects of this image, let us first have a look at its content, and at the message we think it conveys. The object is a typical Renaissance tapestry of the type that has been described as *"the mobile frescoes of the North"* by Delmarcel (1999). The central figure is a crowned lady who points at the sky, and holds a banderole with the inscription *Astronomie*, together with a couple of symbols of ∗ and ○ shape. The second person, perhaps an astronomer, holds an armillary sphere[17] while looking up. The folded lower edge of his robe reveals some letters P, T, O, M,[18] and the sleeve of his robe also carries characters.[19] The *sphaera* displays some zodiac signs – the central one, no doubt, is Aries. A second male person is seated and takes notes, he may be a scribe or a chronicler. That person's robe is also adorned with text characters. In front of him is what looks like a positional observing instrument of the astrolabium type. Four other people (one on a ship) contemplate the heavens with great awe. The landscape clearly is not the Low Countries; the sky shows a waxing gibbous Moon, and more than three dozen stars. Half a dozen stars with tails are present, among which a very conspicuous "wheel of fire" with six fiery outlets.

---

[13]Lemaître (1927): *"v/c mesure donc l'effet Doppler apparent dû à la variation du rayon de l'univers."*

[14]*Röhsska Museet för Konsthantverk och Design*, Göteborg, Sweden.

[15]*circa* 1500–1510, according to Delmarcel & Brosens (2009).

[16]dpi: dots per inch.

[17]Johannes de Sacrobosco (or John of Holywood, 1195–1256) published a standard work on astronomy entitled *Libellus de sphaera*, featuring a similar handheld sphere on the cover.

[18]Perhaps referring to Ptolemy?

[19]Renaissance tapestries often display labels, with a mix of French and Latin words, and Greek symbols, in mixed fonts.

What is the meaning of this scene? Without a caption it is hard to understand all the iconography, but it is very likely that the tapestry is one of a tapestry cycle dedicated to the seven *artes liberales*: music, astronomy, geometry, grammar, dialectic, rhetoric, and arithmetic (Brosens 2000) – although no trace of the other tapestries of such a set was ever found. The celestial phenomenon probably refers to the passage of a comet, or/and to an exceptional meteor shower, or storm. It cannot possibly be the Leonid meteor storm of October 24–26, 1532 because the tapestry was made long before that date. There was a Leonid shower recorded in China on October 24–25, 1498 (see the Table on p. 274 of Littmann 1998), and Rada & Stephenson (1992) mention that in Damascus, on AH 902[20] *". . . star shooting developed and the sky began to clear of clouds. There were around 20 shooting stars in an hour"*. This was about the last exactly-dated shower[21] recorded in both medieval Arabic and European records for about 330 years. The image of the Moon, with an illuminated part of 84%, could perhaps be an accurate time indicator.[22] Ephemeris calculations based on the NASA JET PROPULSION LABORATORY HORIZONS SYSTEMS[23] reveal that on 27 July 1497 the Moon was a waning crescent (9% illuminated), whereas on 25 October 1498 (Gregorian Calendar date 3 November 1498) the Moon was waxing gibbous with an illumination fraction of 77%, which corresponds quite well to the ratio measured on the image.

What, then, does the fiery wheel depicted in Figure 27 represent: a comet? But according to the extensive compilation of Chinese astronomical records *Zhongguo Gudai Tianxiang Jilu Zongji*, compiled under the auspices of Beijing Observatory in 1988, there are no cometary records in Chinese history in AD 1497 or 1498 (Stephenson 2010). Reference to Ho Peng Yoke's (1962) cometary catalogue shows that the comet of AD 1499 was seen[24] in Korea, as well as in China, but not in Europe, and that it is not considered as one of the "great comets".

Or is the celestial iconography only a metaphorization, with the celestial background to be seen as one very special *millefleur* design so typical for Renaissance tapestries? I looked at millefleur compositions at the Paris Cluny museum[25] on many wall hangings from the same period. They are very thematic, with several tapestries displaying cartoon-like sequences, depicting lords and their entourages against an almost abstract background. I perceive one very striking difference between those tapestries and the tapestry discussed here: all other millefleur patterns (various kinds of flowers and birds) are distributed all over the scene, whereas in the case under discussion, the celestial objects are only where they belong: in the sky. And the former never seem to receive any attention from the personages in a tableau, whereas in the latter case (except for the Muse) it is just the opposite. Since all millefleur details at the Cluny museum look very realistic, we might thus be tempted to conclude that the celestial scene is not a millefleur, but an

---

[20]*Anno Hegirae*, or year 902 of the Islamic calendar, corresponding to 27 July 1497 in the Julian Calendar.

[21]Most likely a Perseid shower, see Figure 2 of Rada & Stephenson (1992).

[22]The image on the tapestry is not perfectly symmetric, and shows a quite wide terminator, hence the accuracy of my measurement is not better than 3–5%.

[23]http://ssd.jpl.nasa.gov/horizons.cgi

[24]The comet was visible till 6 September 1499.

[25]*Musée national du Moyen Âge*, http://www.musee-moyenage.fr/ang/index.html

artistic recording of a very special celestial event. In addition, the fiery stars quite well resemble the starlets in Albrecht Dürer's *Apocalypse* woodcuts,[26] made in 1498.

Rests the question whether this rendering of the Moon is to be taken literally or symbolically: the *man-in-the-Moon* image, preferably in crescent phase shown in three-quarter or full profile, was a quite common icon from the 12–13th century on (Montgomery 1999). There is thus no way to know whether the waxing Moon in the tapestry is a symbolic or a high-fidelity image.

What is the use of bringing up this image in the context of this book? There are three facets of publishing related to this discussion, *viz.* the graphical characteristics, the descriptive aspects, and the production features.

1. **The graphics**. The technical aspects of the reproduction in terms of image resolution and color are discussed in Sections 8 and 9. The tapestry itself is, in fact, a very low-resolution object (the physical pixel[27] elements are about 0.46 lines mm^{-1} in the vertical direction with ~ 0.73 pixels mm^{-1} horizontally), photographed and reproduced at very high resolution. Thus, a good-quality reproduction in print will reveal the original pixels, whereas a poor-quality reproduction will blend the original image elements, see the examples in Figure 27. The textual elements inside the art object are somewhat ambiguous: the characters are of different fonts, and some of the symbols even overlap.

2. **The documentation of the original message**. There are many uncertainties in the reconstruction of the artist's undocumented message when only looking at the scene, without any clue of who designed the tapestry, nor when the design was made, nor what was the geographical origin of the scene. Nor do we know whether the phenomena were rendered in an objective way. Any reconstruction based on historical sources remains ambiguous, since the place and time of the observation is not exactly known, and because it is not always possible to convert a given date in the Islamic calendar to an unambiguous modern date format with absolute accuracy. Not to forget that such an analysis is prone to many potential mistakes: authors of historical texts do not always clearly indicate which time system they use,[28] and modern computer applications do not always allow the processing of time in systems other than the Gregorian calendar.

3. **The production**. Renaissance tapestries were created by following a cartoon drawing of the same size (made by an artist-illustrator for a customer or sponsor). Assuming that the client was the person who witnessed the celestial phenomenon, this first step is already beset by some amount of communication noise. Especially since the weaver only could see the back side of the weave, thus the drawing of the cartoon must necessarily have been a mirror image of the finished picture. Delmarcel (1999) mentions that, occasionally, mistakes were made in drawing and coloring

---

[26] Dürer (1471–1528) made a series of 15 woodcuts of scenes from the *Book of Revelation*.

[27] The pixel (picture element) is the smallest discrete component of a (digital) image.

[28] Ho (1962) explicitly flags Gregorian-calendar dates, but does not do so for the Julian-calendar timings.

those cartoons, so that left-handed figures appear, and inscriptions are displayed in reverse.[29] What means that a precise dating based on the Moon's phase may become completely inaccurate if such a reversal would unfortunately have happened. Concerning the intrinsic colors of the artwork, it is most fascinating that medieval weavers already knew how to place threads of different colors so that the eye of the spectator would experience intermediate colors,[30] see Section 9.4.

The foregoing discussion underlines the need for accuracy in any graphical rendering, and in all textual and tabular commentary related to it, as well as in all communications with persons involved in the publishing and production process.

4   Design types of graphs

The first point I wish to clarify is the difference between a plot and a drawing: the former refers to a graphical rendering of tabulated data points, or of data calculated by means of a mathematical function. The latter forms of design are graphs literally drawn using special mouse-driven or algorithm-driven drawing software. A special case of drawings are the schematic diagrams shown in Figures 1 and 2. Text boxes, for example in Section 10 (pages 143 and 150), are also graphical elements (though they are strictly typeset).

More than 50% of the graphs encountered in scientific journals are relational, *i.e.*, the graphs are about data where there is a link between the variables – not just between two quantities, but quite often of multivariate[31] character. Many of these graphical representations are, in one way or another, ordered in series: for example time series (light curves), wavelength series (spectra) or cyclic series (phase diagrams).

In what follows I make a distinction between points/lines and surfaces, a difference that is not related to the true spatial dimension of the graphical content, but to the fact that rendering one-dimensional quantitative information by symbols and lines often yields a different visual impression of the data, than when viewing the same quantities as surfaces.

4.1   Points and lines

4.1.1   Scatter plots

The scatter plot is widely used to show data of two or more related variables. It is particularly useful when the variable of the y-axis[32] is (thought to be) dependent upon the values of the (usually independent) x variable. The data points in a scatter plot are represented

---

[29]The situation is very similar to modern printing, even more so to reproducing photographs from negatives.

[30]Mortimer (1731), on James Le Blon's method of weaving tapestries, states "The Observation of the *compounded Colours* reflected from two Pieces of Silk, of different Colours, placed near together, first gave him the Thought of what Effect of weaving *Threads* of different Colours would be, when all the *Threads* were so fine, as not to be distinguished at a small Distance one from another."

[31]A multivariate system is one whose properties are determined by many independent variables that are individually subject to sampling and measurement errors.

[32]Here, the x and y-axis denote, respectively, the horizontal and vertical axis in the graph.

by symbols that are sometimes joined by straight lines or by curves, see also Section 11.5. This paper includes several scatter plot graphs.

4.1.2   Vector plots

A vector plot is a scatter plot that does not only represent x, y data, but also carries additional information, see for example Figure 4, where the symbols are vectors with directional and magnitude (length of vector) information. Vector plots can get overcrowded when the vector grid is too dense. Note that in Figure 4 there is some overlap between legend[33] and data. This is a pseudo-color image, resulting from converting the gray levels of a monochromatic image to a color image using a lookup table. The shaded band on the right is the color-code scale, which is the key for quantifying the graphical data.

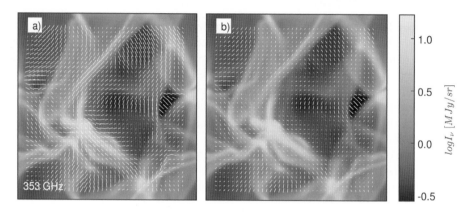

Fig. 4. Simulated polarized dust emission: two models. The scatter plot symbols represent polarization vectors, the colors are log intensity. Source: Pelkonen *et al.* (2009).

4.1.3   Line graphs

Line diagrams are one of the most common tools used to present information. A line graph reveals trends and relationships between data, and illustrates visually how two variables vary with each other.

4.1.4   Time series and phase diagrams

Time series are plots of a quantity (color index, magnitude, velocity, chemical composition, ...) as a function of time. A phase diagram[34] is a time-series curve folded with the

---

[33]The legend of a graph is explained in Section 7.5.1.

[34]Not to confuse with phase diagram in physics, which is a graph showing the limiting conditions for different phases of a substance while undergoing changes in parameters such as pressure and temperature, see also Section 4.3.7.

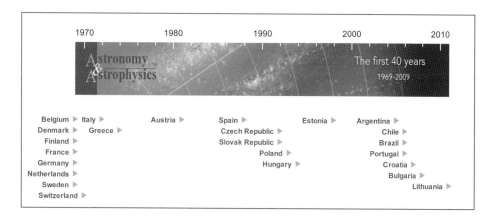

Fig. 5. A&A country membership timeline. The light- and dark-blue tints of labels and symbols are the colors used on the journal's cover (as listed in Table 7).

period of variability, so that the measured quantity is plotted as a function of phase. Phase diagrams are one special way of visualizing cyclic data. Attaching an extra cycle (or part of a cycle) to the time (phase) axis helps the eye to see the cyclic pattern, and avoids the reader to have to jump back and forth between phases 1 and 0. This redundancy is very useful, see the phase axis in Figure 14.

4.1.5 Timelines

A very special time-series graph is the so-called *timeline* – a graphics that conveys a sense of change over time, for example a geologic time scale, or the timelines of evolution or nucleosynthesis. Figure 5 shows some timeline events related to the history of A&A.

Timelines are indispensable for describing large projects and experiments, and for documenting the construction of instruments and telescopes. Time, expressed in Julian Day or civil year (AD or BC)[35] is mostly increasing from left to right along a horizontal x-axis, and can be negative (Fig. 25), but needs not necessarily be linear in scale, nor continuous (Fig. 41). Figure 21 is another example – literally a time line related to the definition of time – including many elements of graphical design discussed in this paper, such as partial grids and omission of the y axis (see also the caption of that Figure). Figure 42 illustrates the use of the cycle number on the x-axis, together with the duplicate time axis in units of year.

4.2 Box-whisker plots

Box-whisker plots display data by showing the minimum of a sample, the lower quartile (Q_1, see also page 153), the median (Q_2), the upper quartile (Q_3), and the highest

---

[35] AD stands for *Anno Domini* and BC for *Before Christ*. BCE can also be used: *Before the Common Era*.

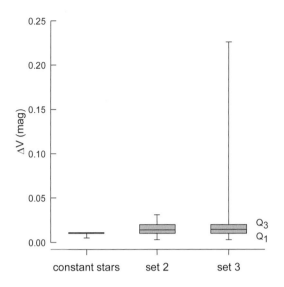

Fig. 6. Box-whisker plot for amplitudes of photometric variability of high-luminosity O and B stars observed by Appenzeller (1972). The sample minimum and maximum are indicated with short horizontal bars, the quartiles are labeled Q_1 and Q_3, the unlabeled horizontal line (Q_2) is the median or second quartile. Set 3 comprises the complete set of amplitudes, set 2 is set 3 *minus* "outlier" R 81, and the first box represents stars with amplitudes ≤ 0.011 ($2\times$ the mean error of one observation). See also Sterken (2011) for an application of box-whisker plots to large datasets.

data point, **without any assumption of the underlying statistical distribution of the data**. Figure 6, for example, is a box-whisker plot that displays the range of photometric amplitudes for measurements obtained by Appenzeller (1972) – *i.e.*, $\Delta_{max}V$ from his Table 1. A further discussion of the use of box-whisker plots to detect outliers is given in Section 11.4, see also Figure 45.

4.3 Surface plots

As stated in the second paragraph of this Section, "surface" here refers to surface aspects applied to the representation of data of any kind, thus not necessarily to three-dimensional data surfaces only.

4.3.1   Histograms

A histogram is a graphical display that shows the frequency distribution of a single variable by dividing the dataset into bins. For each group, a rectangle of base length equal to the range of values in that specific bin is constructed, with the height of each bar corresponding to the count or frequency (or percentage) for the class that it represents. A histogram has an appearance similar to a vertical bar graph (see next Section), but when the variable is continuous, there usually are no gaps between the bars. When the variables are discrete, however, gaps should be left between the bars. One can draw the histogram with adjacent bars touching, or with bars that do not touch – as long as the bin width (for the statistics) does not change. Such histograms are equivalent, and the non-touching version has the look of a bar chart. It is often useful to detach the baseline from the bars, as is done in Figure 28, provided the reader is told how to quantify the vertical scale.

The histogram deals very well with large datasets, and easily reveals the presence of any unusual points or gaps in the data. For good examples, see Figures 5 and 6 of Laurent

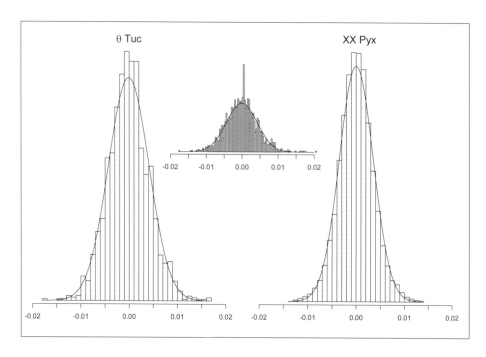

Fig. 7. Histograms of magnitude measurements of θ Tucanae and XX Pyxidis from Sterken (2000). The inset (scale factor 0.75) was made with the same θ Tuc dataset and three times smaller bins.

Cambrésy's paper in Part 1 of these Proceedings, and also Figure 28 below. One glance at a histogram reveals a lot about the statistical properties of the variable: standard deviation, symmetries, outliers (see Sect. 11.4), etc. The histograms in Figure 7 illustrate this very well: they show the distribution of magnitude residuals[36] after removal of all pulsation frequencies from the magnitude time series of two δ Scuti stars. While the XX Pyx residuals support a rather normal distribution with $\sigma = 0.0035$, the θ Tuc residuals are distributed differently, with a standard deviation that exceeds the former by 25%. It is not possible to disentangle the contributions due to intrinsic stellar variability of θ Tuc, and the deviations caused by incomplete frequency solutions or insufficient data content. This example illustrates the assertation of Hoaglin *et al.* (1983) that

AN ANALYSIS OF A DATASET IS NOT COMPLETE WITHOUT A CAREFUL EXAMINATION OF THE RESIDUALS.

The middle graph shows how the visual impression is modified by merely making the bin widths smaller by a factor of three.

Stacked charts (*i.e.*, each bar is divided into relative proportions of more than one variable) are quite popular but, honestly, I always have a problem to get their message at

---

[36]Residuals are what remains after a model has been subtracted from the data: residual = data *minus* fit.

a first glance; I find out that I really have to visually "measure" the relative differences before I can make up my mind. Adjacent-bar histograms do not pose this problem.

Histograms are very often used for inspecting and adjusting the tonal range of an image: they show the distribution of pixels within various intensity ranges, and a properly exposed photograph with full tonality will deliver smooth histogram statistics. Figure 4 covers a complete tonal range with good coverage of all densities from dark to bright, whereas in the reproduction on the cover of this book the dark tones predominate, with a total absence of bright shades, thus more shadows than highlights – as is quite normal for this kind of art object. Figure 43 shows information about each of the color channels in the cover illustration: the left side of each diagram displays the darkest shadows, the right side shows the brightest tones, and the height shows how much of that tone is present. Similar – yet even more extreme – histograms result from astronomical imaging where the numerous dark-sky pixels produce a histogram with typical "underexposed" rendering.

4.3.2   Bar graphs

Bar graphs, also called step plots, usually present data grouped in class intervals, and as such they easily reveal patterns and trends. Bar graphs consist of a horizontal or vertical axis, and a series of labeled bars perpendicular to that axis. The width of each bar is arbitrary, and adjacent bars are often separated by a blank space. The vertical bar graph displays data better than the horizontal bar graph, and is preferred when possible, though horizontal bars perpendicular to the ordinate axis offer the advantage that a horizontal label can easily be added to each bar, see Figure 31 in Paper III.

Bar graphs, like pie charts (Sect. 4.3.3), leave you a free choice in the ordering along the x axis: items may be ordered alphabetically, or ranked according to the magnitude of the depicted quantity, or following any other suitable criterion.

A bar graph and a histogram differ in the following: in a bar graph, frequency is measured by the height of the bar, whereas in a histogram, frequency is measured by the area of the column. Figure 31 in Paper III illustrates the combination of both types of graphs on the same dataset.

4.3.3   Pie charts

A pie chart displays the values of a given variable (*e.g.*, percentage distribution of sub-samples) using a circle divided into segments. The area of each segment is the same proportion of a circle as the subset is of the total dataset and, consequently, the sum of the segments amounts to 100% of the pie – see Figure 29 for a historic example (Newton's color wheel of 1704). The pie segments are often labeled with their actual values, and one or more segments are sometimes separated from the rest of the pie for emphasis. Pie charts with less than a handful of slices should not be constructed, and charts with an excessive number of very thin slices should equally be avoided (see Fig. 8).

Pie charts are very good tools for showing quantitative information when emphasizing ratios of components in a system (see also Sect. 4.3.7), but the viewer of such graphs should be aware that, when comparing quantities visualized by surfaces, the perception of surface will lead to overestimation of the real ratio. Figure 8 shows the relation be-tween the percentages and the corresponding surface areas (linked by a parabolic curve): whereas the ratio of the percentage levels of the two major segments is about 1.3, the ratio of surface areas is 1.6, which leads to a perceptual misrepresentation that worsens with increasing ratios. The leftmost inset replicates Figure 2 of Bertout's paper in Part 1 of these Proceedings, with area color codes "picked" from the original postscript file (see also page 139 on color picking).

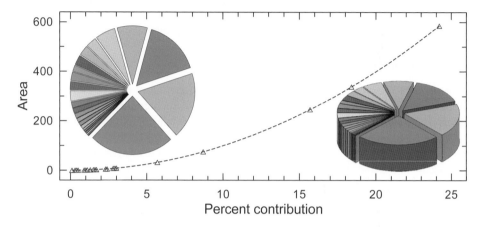

Fig. 8. Submissions from A&A sponsoring countries in 2010 in two pie chart graphs. The line is the parabolic fit of percentage contributions *versus* area for the left pie chart.

Pie charts (and also bar charts) are often presented in pseudo 3-D by adding "depth" through a *lip* of substantial thickness. Adding such an extension (thus using volume for representing a quantity, see the rightmost inset in Fig. 8) will enhance the deceptive per-ception even more, especially when the pie with the largest component is placed in front. Note that both insets have the same linear dimension along the *x*-axis. The explosion factor provides a different viewing aspect in 2-D and in 3-D: the three-dimensional con-cept takes less space, but produces higher distortion, *i.e.*, it increases the *lie factor* (see Sect. 11.3).

The only real difference between the original graph and Figure 8 is that in the former the variables are ordered alphabetically, whereas in the latter they are aligned according to rank order of increasing contribution. This pushes all small slices together, and illustrates the point made already that inclusion of many thin slices should be avoided. Small slices are notoriously difficult to label without overlap or without distracting the eye with an overdosis of labels. The original graph is a good example of proper labeling: labels and the element that is labeled (symbol, area or line) are linked, and should thus be close in space. Note also that the smaller the slices, the larger must be the *explosion factor*, *i.e.*, the amount of eccentricity added at the center of each slice so that their central parts do not stick together.

4.3.4 Bubble graphs and pictographs

Bubble graphs are similar to scatter plots, but they display numerical quantities with plot symbols that have a surface area proportional to the value of the variable. They are also two-dimensional representations of one-dimensional data. Bubble graphs are very popular for making starmaps, but – like pictographs – they are known to misrepresent the data, because the visual sensation through surface rendering is not a linear function of the original quantity.

Pictographs use picture symbols to display statistical information, and are used most often by professional graphic designers. Pictographs should be used with care, because the graphs may, either accidentally or deliberately, misrepresent the data quantities.

4.3.5 Contour maps

Contour maps (also called isolines or isograms) of a function of two variables are a set of lines, along which the depicted function is constant. Such graphs allow visualization of extremely high data densities, provided that an appropriate contour interval is selected: too close contours may occasionally blend. The gradient of the function is perpendicular to the contour line, and lines are often numerically labeled. Labeling is mandatory when the spacing of the lines is uneven. Contour maps are ideal additions to pseudocolor maps, but only when an explanation of the coloring conventions is supplied (in the caption, or as a legend).

4.3.6 Sky maps

Sky maps are projections, and the most common type in astronomy is where declination or latitude (galactic, heliocentric, …) goes along the y-axis and right ascension or longitude runs along the x-axis. All projections distort, in one way or another, hence the map designer should be vigilant about aspects of conformity, *i.e.*, distance, azimuth angle, scale, and area conservation.

An example of a very handsome and versatile sky map is the annual *Skygazer's Almanac* in *Sky & Telescope*, which is a handy two-page foldout chart that summarizes the rising and setting times of the Sun, Moon and planets. Each chart depicts multiple time-lines with celestial events plotted on an hourglass shape, showing in well-chosen colors when celestial cyclic phenomena are observable.

4.3.7 Ternary diagrams

Two-dimensional ternary[37] diagrams (also known as triangular diagrams) show relative proportions (in percentages) of compositional data[38] in a three-component system. Each component is normalized on a scale of 100%, and the coordinates are called barycentric coordinates. Points outside the triangle have at least one negative coordinate (see Fig. 30). Ternary diagrams are thus used for displaying 3-D data in a 2-D graph with a triangular

---

[37]The adjective ternary means "having three components or elements".

[38]Not to confuse with *composite data types*, which are data types constructed in a computer program.

coordinate system. These graphs are, in some sense, comparable to pie charts, where the components also carry only relative information. There are three reasons to include ternary diagrams in this discussion:

1. they are used to illustrate the relative percentages of three-color primaries in colorimetry (see Sect. 9.5.1);

2. they are (for astronomers) relatively unfamiliar, thus need special attention when applied or read, and

3. they are very useful for supporting non-quantitative arguments, see Figure 36, and also Figure 13 in Paper III.

Each corner of the composition holds 100% of the corresponding component, and all mixtures of two components lie on the edge of the triangle that opposes the apex of the missing component. The closer a point lies to an apex, the more of that component is in the mixture. It is quite common to label the three apexes, but the percentages of the whole, or the normalized scales (0–1), are mostly not given. Users of ternary graphing software should verify that autoscaling and normalization occurs on the full data range, and not only on the range of the data in the plot.

Figure 30, for example, displays two versions of Maxwell's color space,[39] also called *Maxwell's triangle* – one of the very early ternary diagrams. Figure 33 also illustrates the principle of ternary plots, though not with an equilateral triangle, but via transformation to cartesian-type x, y axes where each x, y contains only relative information through the implicit condition $x + y + z = 1$.

Ternary diagrams are not easy to read and to interpret, moreover they are not invariant to some coordinate transformations – for example, adding a constant in order to render all coordinate values positive transforms the data content of the graph. Some basic statistical quantities (variance and correlation) are not invariant, since most statistical techniques are designed for free-ranging data, and not for bounded variables in ternary visualizations. Patterns (contour lines) in such diagrams may thus have a different meaning, and spurious correlations may result, Euclidean rulers will shrink when approaching the edges,[40] grid lines can change, and this all has an impact on outlier metrics (see Sect. 11.4). The triangular inset in Figure 33 accentuates not only distortions of the shape of embedded curves, but also illustrates the mental switch needed to read off coordinates along the triangle's sides.

4.4 Genuine 3-D surfaces

Genuine three-dimensional surfaces (for example, point-spread functions and 3-D models) need real 3-D visualization. Figure 9 shows an example of a surface plot of the distribution in solar latitude of the total sunspot area as a function of time (for the time interval 1877–1902). Three-dimensional graphs have very specific elements that need attention, *viz.*:

---

[39]Color models and color spaces are explained in Section 9.4.

[40]Distances between pairs near the center and along edges may seem equal, but are not, see Figure 36.

- contrary to what is said in Section 7.5.4, grid lines can be most useful when applied as a surface mesh, since mesh plots are ideal for visualizing, though they are much less suitable for precise quantization – grid density is thus an important parameter: if the density is too coarse, the rendering of the surface becomes inadequate;

- it may happen that some sides of a 3-D surface are hidden by what is in front, or that a data aspect is obscured by an axis, see Figure 9, where the highest peak almost coincides with the vertical axis at the back;

- an important visual facet is the *lighting angle* (45° in this case) that produces the shadows: a change in this angle can bring forth a totally different view;

- *depth*, the third dimension shown at a selected angle (solar latitude in Fig. 9), also strongly influences perception, and affects legibility of axis labels;

- opaque background walls may lead to uncontrolled clipping of data and outliers, although this is not the case in Figure 9, since there are no data beyond the axes limits;

- comparison of both graphs in Figure 9 shows that replacing missing values by zero-values produces a totally different perception, and even adds a spurious high-frequency signal.[41]

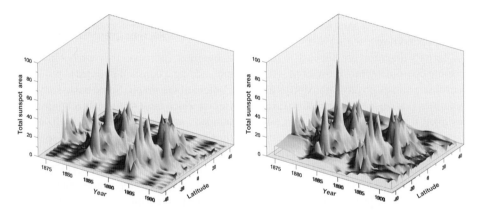

Fig. 9. Sunspot data from Maunder (1904). *Left*: the missing values from Maunder's Table were included as zeros. *Right*: based on the original Table without replacing the missing values with 0.

The dataset shown in both panels of Figure 9 is a very famous one, which led to the well-known *Maunder butterfly diagram* that demonstrates that the latitude of sunspot occurrence varies cyclically on a time scale of about 11 years, with sunspots forming at high latitudes during the minimum of the cycle, and getting closer to the solar equator towards the cycle maximum. And not only the data are crucial, also the initial Maunder

---

[41]Replacing missing values by zero-values in the wings of a histogram may yield a totally spurious best-fit theoretical distribution function.

graph was fascinating, even the story of the publication of the original data, and the subsequent graphical renderings by two different persons leading to contradictory conclusions, is most intriguing, and deserves some explanation.

Maunder's (1904) paper commences with the following statement: *"Rather more than a year ago the Astronomer Royal communicated to this Society a paper which he had desired me to prepare".*[42] That paper summarized, in graphical and in tabular form, the results of more than 30 years of work in the measurement and reduction of about 9 000 photographs of the Sun, involving more than 250 000 measurements made in duplicate. The data comprise the total area of whole sunspots divided by the number of days of observation, expressed in millionths of the Sun's visible hemisphere, for each degree of solar latitude. That paper discussed the data with graphics consisting only of bar charts of the annual spot area as a function of heliographic latitude. These graphs showed gross trends, however failed to reveal any striking patterns.

Surprisingly, one year later Maunder supplements this very paper with "other diagrams",[43] together with new diagrams that he deemed necessary for refuting some conclusions – entirely based on Maunder's own data – since published by William J. S. Lockyer. He refers to his own "authorless" paper[44] as *"the Greenwich paper of a year ago"*. Two aspects are of interest in this discussion:

1. the somewhat strange way of publishing such a large and original dataset without author, and

2. why Lockyer (1903) and Maunder (1904) come to **different graphical interpretations of the same data**.

Not to speak of Lockyer – who very well knew who had composed the 1903 data paper on sunspots – who refers to it in terms of *"the fact that the Astronomer Royal has quite recently published the values of the amount of sunspot area for each degree of latitude for each hemisphere ..."*, without even mentioning the composer[45] of that paper.

Maunder (1904) presents curves of distributions of spotted areas in zones of 5° latitude, and states that his Figures are "practically" the same as those of Plate 15 in the anonymous 1903 paper, but that the new presentation in 5° zones describes the facts better than the more detailed (bar-)graphs given previously. Indeed, the 1°-resolution graphs are dominated by isolated outbursts and their features are, as he describes it *"a feature of solar book-keeping, not of solar physics."* In other words, **he describes the positive effect of binning in bringing out salient features – and thus emphasizes the impact of binning on the perception of the graphical content.**

Maunder (1904) questions whether Lockyer's year-to-year joining of spot-centers to form "spot-activity tracks" (Fig. 10) is justified, and he regards the linking together of *"certain selected"* spot-centers as *"a matter of personal predilection, not as one of solar physics"* (see also the discussion on dot joining in Sect. 11.5). He then illustrates the

---

[42] This paper is MNRAS 63, 452. The Astronomer Royal in 1904 was William H. M. Christie (1845–1922).

[43] Diagrams, he says, he composed at the time when he prepared the 1903 paper for the Astronomer Royal.

[44] ADS refers to this paper as authored by N/A (Bibcode 1903MNRAS..63..452), see Paper I, Section 9.2.

[45] Here I use *composer* because the paper was not properly authored.

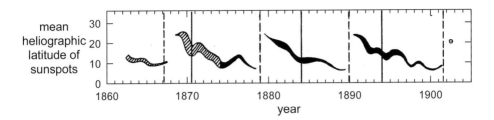

Fig. 10. Mean daily area of sunspots, Lockyer (1903) Plate 1. The caption of the original Figure does not specify the meaning of the different patterns of shading.

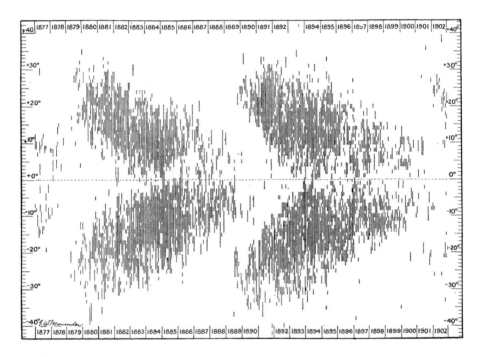

Fig. 11. Plate 16 of Maunder (1904): distribution of spot-centers in latitude, rotation by rotation, of the Sun (not on the same scales as Fig. 10). Compare with Figure 52. Note that the time axis labels for 1891 and 1893 were omitted in order to avoid overplotting with data.

arbitrary character of Lockyer's sunspot-group associations by showing the distribution in latitude of spot-centers in the northern hemisphere, and finally presents his ultimate diagram in "Plate 16": the distribution of spot-centers in latitude, irrespective of their surface area, rotation by rotation of the Sun, covering a quarter of a century, *i.e.*, two solar cycles (Fig. 11). This visually convincing diagram is now known as his famous *butterfly diagram*. Figure 12 shows a modern version of this diagram, together with a graph that basically comprises the same data, with the spatial component removed by averaging (lower panel in Fig. 12). The Table below lists the principal differences between the modern and the vintage versions of these diagrams.[46]

Item	Maunder (1904)	NASA butterfly diagram
Ordinate scale	linear in solar latitude φ	linear in $\sin\varphi$
Data extent	restricted to $-45° \leq \varphi \leq +45°$	full range $-90° \leq \varphi \leq +90°$, including dataless regions
Dimension	2-D	3-D by means of red and yellow color coding
Database	contains no 0-values	contains zeros, which are rendered as white pixels

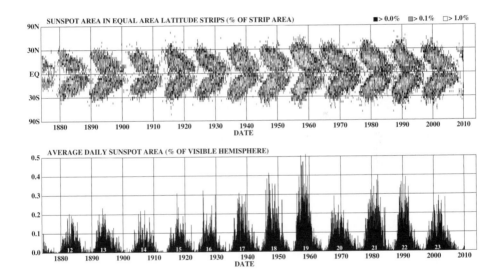

Fig. 12. Daily sunspot area averaged over individual solar rotations. *Top*: sunspot area in equal-area latitude strips; *bottom*: average daily sunspot area (in units of 0.001 visible solar hemisphere). The white labels 12–23 in the lower diagram indicate the solar-cycle number. Source: David Hathaway, NASA/MSFC.

4.5 Drawings and cartoons

Drawings and cartoons are often utilized to demonstrate a scheme or plan, or a model. Figure 13, for example, shows the family tree of traditional photometric systems published by Sterken (1992). This graphical metaphor clearly suggests that photometric systems result from pruning and grafting the tree – in other words, it illustrates that

---

[46]Note that this tabular structure is included, intentionally, without Table number.

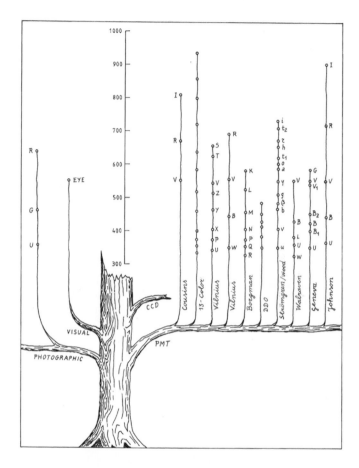

Fig. 13. Genealogy tree of classic photometric systems. Each main branch symbolizes a photometric detector, and the buds represent photometric bands arranged by wavelength along the vertical axis. Reprinted from *Vistas in Astronomy*, Vol. 35, *"On the future of existing photometric systems"*, C. Sterken (1992), with permission from ELSEVIER.

maintenance of a standard system is a prerequisite. This chart displays no strict quantitative information, although the vertical wavelength scale is correct. There is no need for an *x*-axis in this graph, since the layout of the horizontal axis is arbitrary, and the ranking of the items along the *x*-axis has no meaning, except for a purely aesthetic arrangement.

Note that the caption of Figure 13 gives the full bibliographic identification of the source, as requested by copyrightholder ELSEVIER. This reuse was free of charge, although it was required to obtain a license through RIGHTSLINK,[47] and sign a clause of compliance with the license terms and conditions.

4.6   Which design type of graph?

The Maunder–Lockyer story, explained in the previous Section, clearly illustrates that different graphical representations of the same dataset can very well change the viewer's

---

[47]http://www.copyright.com/search.do?operation=show&page=ppu

perception, and hence also the viewer's analysis and conclusions. Different kinds of graphs may thus have different powers of convincing the reader. So, before embarking on any graphical design, it is good practice to spend a moment of reflection about which type of graph will be used. This question is thus:

> **the data being equal, which graphical design conveys my message best?**

The first and foremost answer is that you must try to avoid complexity by selecting the proper type of graph. Take, as an example, the phase diagram in Figure 14, the left part of which is reproduced from van Genderen & Sterken (2007). The manuscript submitted for publication, however, originally contained the right diagram of Figure 14 to illustrate a small modulation ($\sim 3\%$) of the optical brightness around phase zero of the very eccentric ($e \sim 0.5$) binary BP Cru, but the Editor of the journal could not be convinced. Replacing the bar graph with the scatter plot finally made our statement more acceptable. The only reason why we had opted for a bar-graph approach in the first place, was to enable direct comparison with similar diagrams in the literature (see van Genderen & Sterken 2007 and references therein). This example also shows that designs of graphs are copied or imitated from one source to the other, and as such get consolidated, unless "alien eyes" force the designer to reconsider the concept. The same applies to tri-variate data and ternary diagrams: they do not always provide the best mental impression to the eye untrained for their interpretation.

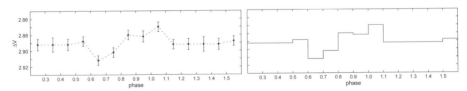

Fig. 14. The same data presented in different graphical forms. Source: van Genderen & Sterken (2007). Note how the labels and line widths suffer from the reduction factor.

Another element of design to consider is the aspect of symmetry: whenever part of a graph is a mirror image of another part, you may opt to omit the symmetric portion, because erasing redundant parts increases the information content rather than reducing it. That is, provided that you tell the reader that there is symmetry involved. A typical example of symmetric graphs is found in spectral-window frequency curves in Fourier analysis. Beware, though, that a symmetrical distribution on a log scale implies an asymmetric distribution of the argument variable.

5 Computer graphics

There are two basic types of computer images: vector images and raster (bitmap) images. The former are best suited for drawings (like scientific graphs), the latter for photographs, paintings, and image processing. Graphics files can contain raster information, vector information, or both.

5.1   Vector images

Vector graphics are images that are described by vector objects, *i.e.*, they are completely defined by points and curves (vectors), and by instructions (line type, line thickness, and color) for the vector-based drawing software to produce lines and curves by connecting the individual points (either with straight lines or with *Bézier curves*).[48] Well-designed vector drawings are resolution-independent and can be scaled to the maximum resolution of the printing device without any loss in sharpness. In other words: **vector graphics are scale (*i.e.*, output device resolution) independent**. Vector images are ideal for designing logos and other objects that have to be frequently resized, and because vector images only contain the data needed to describe the vectors, their file sizes can be quite modest, as long as the graphs are not too complicated.

5.2   Raster or bitmap images

A raster or bitmap image is a matrix of pixels with, *for every cell*, the instruction that describes its pixel content: examples of bitmaps are scanned photographic images and digital pictures. Bitmaps can take up a lot of disk space: an uncompressed good-resolution full-page color picture can easily amount to 50–100 Mb. Such images are resolution-dependent, therefore embedded text cannot be correctly scaled. There are basically four main types of bitmap images:

1. *line art*: contains only 2 colors, mostly black and white (these images are named bitmaps because only 1 bit [0 or 1] is used to describe each pixel);

2. *duotones*: contain shades of only two colors, often black and a second color;

3. *grayscale images*: contain shades of gray, but also pure black and white;

4. *multitone and full color images*: either tritone (three colors) or full-color (quadri).

An important term is *bit depth*, which is the number of bits used to store information about each pixel in the image. The higher the bit depth, the more levels of gray or color shading that can be rendered. Monochrome images are 1-bit images (each bit contains two levels of gray), grayscale images are 8-bit images. A four-color image usually is a 32-bit image with eight bits of data (2^8 or 256 levels of shade) per color channel,[49] yielding 16.7 million colors.

The term bit depth mostly refers to the number of bits per pixel in the image file, but can as well refer to the largest output number that a digital camera can produce. That number depends on the capabilities of the analog-to-digital signal converter. A 16-bit converter provides 65,536 levels of information, a 12-bit converter divides the signal into

---

[48]Bézier curves are segments resulting from cubic interpolations between control points, and have the property to preserve their smooth appearances, even at high magnification.

[49]Color channels are explained in Section 9.4.

4096 levels (although the 12-bit information is often stored in a 16-bit image format). Most CCD cameras use a 16-bit format for capturing and saving (for example in FITS format, see Sect. 5.4), and many digital cameras have settings that allow images to be shot and stored in the very flexible 16-bit RAW format (or in 16-bit and 8-bit JPEG as well). Note that there does not exist a common RAW format among camera types and brands, since each camera manufacturer has established its own proprietary RAW standard (NEF by Nikon, CRW by Canon, etc.).

5.3 Converting bitmap to vector data and vice versa

Transformation of images from raster to vector data is sometimes needed when scanning a photograph of a logo or a drawing (like Fig. 13), because such graphics have awkward magnification and file size properties. The transformation procedure to vector curves is also called *tracing*, a process comparable to *Optical Character Recognition* (OCR) of text, and results in lines and characters to be recognized and saved in vector format.

Vice versa, vector drawings must be converted to bitmaps for use on web pages, and many drawing-software packages allow saving vector data as bitmap files via the Export menu. Image-handling software can import some vector file formats through *rasterization*, and let you save the result as a bitmap.

In both kinds of applications, the user should not forget to specify the resolution and mode (grayscale or color) of the output bitmap, as well as the data-compressing method.

5.4 File formats

The reader (human or machine) needs to know how the data are stored in an image file. This information can be quite complex, and is called *metadata*, *i.e.*, "special data about data". Appropriate software is needed to reconstruct vector graphs, though in principle visualization of bitmaps is simpler – except for the data compression and all forms of included image previews.

Since image files are written as output of a computer application, and then read again by another application, this process very much follows Shannon's model of communication (see Sect. 2). Some elements related to file formats thus contribute to confusion and communication noise:

1. there is the *native format* designed by the developer of the software;
2. formats, by design, are made to deal with specific information, and are not necessarily transformable;
3. formats change with time (via higher software versions);
4. formats have been shown to contain programming bugs and hidden information – even malicious code containing safety hazards;
5. format is not the only aspect to consider: not all image-writing applications allow the user to specify the desired resolution, a problem when outputting information;
6. the perpetual problem of transporting data between operating-system platforms;
7. some formats are open standards, others are not, and

8. the data compression: some data-compression methods destroy some of the data during compression (so-called *lossy* compression, where the amount of compression is highly dependent on the image content), and thus modify the information content of the image.

So, if transforming from one format into another is such a problem, why then do we do it in the first place? There are several reasons, the first of which relates to making the graphs, and the second concerns getting the graphics published:

1. file conversion is needed if you collaborate with colleagues who use other software – even a previous version of the same software, and

2. bitmap and vector graphs can be saved in a wide variety of file formats, but very few of these are suitable for *prepress*[50] operations: Publishers are not prepared to deal with a wide variety of formats, and quite often limit your choice to TIFF or EPS.

I have a recent example of the first point: in one project, a participant sent us CCD frames in the *Pictor 416* FITS format, which could not be read with standard FITS readers (IDL[51] or SAOIMAGE[52]). But the frames could be handled by MAXIM[53], which in turn produces output in a FITS standard that is compatible with IDL and SAOIMAGE.

Table 1. Principal formats used in the scientific community, with file extension and bit depth per channel.

Ext	bit	Description
BMP	8	format that is not suitable for prepress
EPS	8	format that can contain bitmap as well as vector data
FITS	16	format for data capture, data reduction and analysis, not suitable for prepress
GIF	4–8	mainly used for internet graphics (so-called indexed color)
JPEG	8	applies a lossy encoding method, and is mainly used for internet applications
PDF	8	versatile file format that can contain any type of data
PICT	8	format that can contain both bitmap and vector data, mainly used on Macintosh computers
PNG	8	image format that employs lossless data compression, also used for web applications
RAW	16	image processing, not suitable for prepress (every manufacturer has own RAW standard)
TIFF	8–16	the best bitmap file format for prepress and other non-web applications

Table 1 is a concise overview of the principal formats in use in the scientific community, and the subsequent list gives some more detailed information about each type.

- BMP (*Windows bitmap format*): a Windows file format, ideal for Windows adepts as long as no transfers to other platforms are required.

---

[50]The *prepress* phase includes all processes and procedures that occur between the creation of an image and the final printing.
[51]Interactive Data Language.
[52]http://fits.gsfc.nasa.gov/fits_viewer.html
[53]http://www.cyanogen.com/

- EPS (*Encapsulated Postscript*): developed by ADOBE, and tuned for including graphics in typeset documents, see also Paper I.

- FITS (*Flexible Image Transport System*): the file header is in ASCII, the body contains graphics information in binary format. The FITS format was developed for exchanging scientific datasets, and it is commonly used to store astronomical photographs. It supports both 8-bit and 16-bit black and white images, but there is currently no standard for encoding color images.

- GIF (*Graphics Interchange Format*): used to place images on the Web. GIF uses *indexed color*, *i.e.*, each pixel's data only consists of an index number that identifies one of 16 (4 bit) or 256 (8 bit) selected colors of a user-defined *palette*[54] that is embedded in the file.

- JPEG (*Joint Photographic Experts Group*): a standard image file format for multiple computing platforms, allowing for variable levels of (always lossy) compression. JPEG always discards information that cannot later be recovered: saving and reloading a JPEG color image will always degrade its color information. JPEG is excellent for archiving images, *i.e.*, images that need no additional processing. JPEG2000 is a format similar to JPEG, though better because it uses a wavelet compression method and has a lossless option.

- PDF (*Portable Document Format*): an open standard for document exchange created by ADOBE Systems, used for documents, and characterized by portability (reliable viewing and printing across platforms).[55]

- PICT (*Apple Picture Format*): 8-bit format that can contain both bitmap and vector data, mainly used on Macintosh computers.

- PNG (*Portable Network Graphics*): was developed as a replacement for GIF. It supports all image types, and works across different platforms, and features lossless compression.

- RAW: contains the original pixel information directly from the detector, untouched by lossy compression and demosaicing[56] – in fact, the FITS format is a RAW format.

- TIFF (*Tagged Image File Format*): a very versatile and transportable format that is supported across many different platforms. It is a good choice, both for image file storage and for exchanging images with Macintosh, Windows or Linux-operated computers, but TIFF files are considerably larger than JPEG files. TIFF has many variants – some of which utilize lossy or lossless compression – and the user can occasionally suffer from incompatibilities between files generated by different applications. TIFF supports both 8-bit and 16-bit per channel images.

Submission to a Publisher in any format other than TIFF or EPS is not recommended, because these are the standard graphics file formats compatible with the composition software in the printing industry. JPEG is acceptable only if saved with the least compression settings (*i.e.*, highest quality).

---

[54]The concept *color palette* is described in Section 9.4.

[55]Invented by ADOBE, but not owned by ADOBE: the format is not proprietary – it belongs to the community.

[56]The processing pipeline that builds a color image from the outputs of the color-filtered detector pixels.

Note that all formats in Table 1 are ≤8-bit formats, except FITS, TIFF and RAW, which contain 16-bit information. The transition from 16-bit to 8-bit – if necessary – should be the last step in the image processing, *i.e.*, all modifications (color, brightness, cropping, contrast) of 16-bit images should be done in the 16-bit environment before saving in 8-bit. Converting 8-bit to 16-bit makes no sense because adding the extra bit depth does not add any information: ***image storage depth* is not to be confused with *image information depth*** (the format cannot add, but for sure it can destroy information content). Once a file is damaged by saving as JPEG, the harm can never be undone by converting it back to TIFF or PNG!

6 Photographs

Journals reproduce photographic images, but what you see in print are not photographs, but so-called *halftones*. In printer's jargon, the term halftone can also refer to the image that is produced by this reprographic process, so if a Publisher asks you how many halftones are in your paper, then the number of photographic images is meant. Another confusing misconception is the reference to color versus black-and-white photographs. We almost never deal with photographs that only contain black and white: most photographs are grayscale renderings. A real black-and-white image is Figure 16, which is a 2-bit composition: each pixel is either black, or white (0 or 1), without graytones (that have bit values 0–255).

6.1 Halftones

Photographs (but also images on computer, TV and movie screens) are continuous-tone images including a wide range of colors or graytones. But XEROX copiers and laser or inkjet printers only have black ink or toner – no shades of gray (thus 2-bit). For printing such an image, it must be rendered into a halftone by a reprographic technique – *the halftoning process* – that simulates the photograph by rendering a binary image of equally-spaced dots of one color (also called halftone cells) and of varying size and density, which create the illusion of variations of gray or continuous color.[57] The rendition fools the eye-brain detector into the perception of shades by using the white of the paper and the color (or black) of the toner.

In desktop publishing, halftones are automatically generated by the printing device when a file is sent for printing, and the halftoning parameters are by default contained in the postscript printer setup file.[58] The most important parameter is the *halftone screen frequency*, which is measured in lines per inch (lpi is the printing industry's term for resolution) and denotes the number of dots that fit in one inch. Note that the quality of the printing paper also plays a role: very fine halftone cells are of no use on low-quality porous "newspaper" paper, since the small cells will "bleed" and produce clumps

---

[57] In the past, photographs were reshot through a halftone screen, nowadays the process is entirely digital.
[58] File with a .ppd or .PPD extension.

of ink. Newspapers use 85 lpi, but journals and magazines use 133–175 lpi.[59] The line screens used by the printer of A&A are 170 lpi for the cover, 133 lpi for articles printed in black (on a rotary offset press), and 150 lpi for articles and pages printed in color (on a sheet offset press). For a four-color image (also specified as *quadri process*), four halftone screens at different screen angle are applied – one for each ink used in the printing process. A second parameter is the *halftone screen angle*: a typical sequence can be 45°, 75°, 90° and 105° for each of the colors (see Fig. 38).

As the lpi parameter comes into action only at the moment of printing, the result will primarily depend on the image resolution that was delivered (see also Sect. 8). The rule of thumb is to ensure that the delivered image should have a resolution of at least the resolution at the output side of the printing process. In other words:

GARBAGE GRAPHICS IN = GARBAGE GRAPHICS OUT = ALWAYS!

QUALITY GRAPHICS IN = QUALITY GRAPHICS OUT = SOMETIMES!

We are used to saying that a photograph is *taken*, but the correct technical term is *captured*. Any artwork image based on the photographic process is first captured, and then *rendered*. There are two types of rendering: let me call the first the *realistic* one, the second the *artistic* variant:

1. the former simply is the standard process of cropping and trimming an image, with unsophisticated optimization of contrast and color;

2. the latter is any kind of rendering that changes the straightforward interpretation of the graphical content through trimming of unfavorable picture elements, drastic changes of contrast, hue and saturation – in short any form of interpretive rendering that willingly or unwillingly changes the basic data content (see also Sect. 11, and Paper III).

6.2   Halftones and Moiré patterns

Moiré patterns are seen when a repetitive structure is overlaid with another such structure, with the line elements nearly superimposed. When scanning a photograph (*i.e.*, halftone) from a book or journal, Moiré patterns are produced by interference between the scanner sampling grid and the halftone screen of the original image. Such patterns may also appear when scanning a painting, because the line- and column pattern of the canvas will interfere with the scanner's sampling grid. All scanners will do this, for each of the color screens involved. For example, a 150 lpi halftone scanned at 600 dpi will result in a periodic 200 dpi Moiré pattern. To avoid any interference caused by the grids, line screens must be about 15° angles apart. Figure 15 illustrates the effect in two images of a polar-coordinate grid (with a line pattern of about 25 lpi) acquired with the same flatbed

---

[59] You obviously need not worry about paper quality when writing for a journal or for a proceedings Publisher, but you have to consider this aspect when producing a PhD-thesis manuscript at a local printshop!

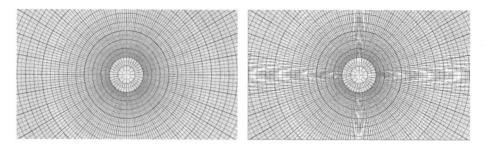

Fig. 15. *Left*: direct print of scanned polar-coordinate grid; *right*: image viewed on monitor screen.

scanner. The left image is the direct print of the scan, and the right image is a screenshot of the scan as it is rendered on the monitor screen[60] showing, respectively, mild and strong Moiré interferences. If you scan halftones, you must consequently process the result with a *descreen filter*. Moiré patterns also appear in printing when the halftone screen angles of two colors are less than 30° apart.

7   Elements of a graphical image

Graphs must be effective, simple, clear, and aesthetic. That means that each of the composing elements of a graph must be worked out with proper care: neglecting one of the elements may very well destroy the good look of an entire composition. Moreover, elements in a graph strongly interact with each other, and this interaction may change the aesthetic appearance, even the scientific message. Pure text as graphics also occurs, for example in Figure 4 in Paper I, but we have seen that graphics without any words are meaningless, hence an important issue is the typesetting of textual components, in the first place because the typeface in a graph is often different from the typeface in the body of the text.

All good graphics contain textual characters, some of which are added by you, the designer of the graph: axis labels, graph titles, data labels and legends. Many of the components of graphical images discussed below are automatically generated by the drawing software. Even if you master the settings and options of your graphics software, it is still useful to verify and adjust these elements, so they can optimally serve the reader. It is, therefore, important to explain a few basic facts about typography.

---

[60]MLCD 20″ monitor with 32-bit "highest" video setting, *i.e.*, 1600 × 1200.

7.1 Aspects of textual components in graphics

7.1.1 Typefaces and fonts

Typefaces and fonts also belong to non-data parts of a graph. *Typeface* refers to the visual design characteristics of a specific character type. A *font*[61] is a set of letters and symbols of a specific design. A font simply is a variant design of a typeface: a text (for example a letter) that comprises only one typeface can still contain many fonts. One example is the classical and popular typeface Helvetica (a modified version of Helvetica became a Windows font called Arial[62]). So if the typeface is Helvetica, the font could be Helvetica 12pt, containing all the letters, punctuation marks and symbols needed for typesetting in typeface Helvetica at 12 pt size.[63]   In the past, fonts were sets of metal shapes of a particular typeface; nowadays fonts are files on a computer that describe the typeface.[64] Although font and typeface have different meanings, the terms are often interchanged in the context of digital printing. The more so because digital fonts are resizable, hence size-specific font nomenclature referring to point size has become irrelevant. The correct phrasing is to speak of *choosing a typeface*, but *installing a font*.

Fixed-resolution bitmap fonts do not print or display very well on some devices, hence authors are advised to always use scalable fonts (also called *outline fonts*). Scalable fonts, when sent to a screen or to a printer, are transformed to dots at the output device, a transformation that is called *rasterization*.

There were originally many different formats for digital fonts, fortunately some standardization has developed, see Table 2. One of the first standards was Adobe's Type 3 bitmapped fonts, which were *open source*[65] from the beginning, and which were destined for the publishing and graphic arts industry. Adobe then developed outline Type 1 fonts – which included *font hinting*[66] – but the company did not disclose the source code. Apple developed its own TrueType scalable font technology in the 1980s, and TrueType fonts were built into the operating systems of Apple and Windows computers. Type 1 or TrueType are scalable vector-based fonts (see Table 2).

For documents using Type 1 fonts, the Adobe Type Manager program will automatically create bitmapped fonts of any size for display without embedding the fonts in the pdf document. Type 1 Fonts are implied by specifying Adobe Times Roman, Courier, Helvetica, and Symbol PostScript fonts for the entire paper, *including all graphics*.[67] Problems can arise when using PostScript Type 3 fonts – for example the Computer Modern Roman font family used by LaTeX – that are written to a less strict syntax than the Type 1 fonts. The same applies when annotating photographs. It is the task of the

---

[61] The word *font* comes from French *fonte* (casting) or Latin *fundere* (to cast).

[62] A non-standard postscript font that may cause problems with specific dvi viewers.

[63] Section 7.1.2 explains the *point*, a typographic unit.

[64] Metal fonts were supplied in a case for each font (like Helvetica 8pt Roman, Courier 12pt Italics). Digital fonts are contained in directories – sometimes called *font suitcases* – and when typesetting in LaTeX, the directory structure should follow the TeX Directory Structure, see www.tug.org/fonts/fontinstall.html

[65] *Open source* describes practices in production and development that promote access to the end product's source materials (Wikipedia).

[66] Hinting is mathematically adjusting the display of an outline font so that it lines up with a rasterized grid.

[67] Adobe Acrobat Reader's Document Properties/Fonts will display a list of fonts used in the pdf file.

Publisher's copy or production editor to look out for such complications (see also the paper by Agnès Henri in Part 1 of these Proceedings).

Table 2. Principal "open" and multi-platform font standards used in typesetting.

Type	Description	Device
PostScript Type 3	Bitmap or outline fonts, open source	Print only
PostScript Type 1	Outline font description of characters by means of points and curves. Resolution independent and scalable.	Print only
TrueType (TT)	Outline font designed as a competitor to Type 1 fonts used in PostScript format	Print and Screen
OpenType (OT)	A cross between Type1 and Truetype.	Print and Screen

7.1.2   Typesetting units of length: pixels, points and picas

In typesetting, and in graphics, there are several typical units of length. One class comprises the absolute units of length: centimeters, millimeters, inches, points, and picas. The other group includes relative units such as ems,[68] pixels, and percentages. Absolute units of length are used for printed media, and relative units are used for on-screen reproduction. The *point* is the smallest unit, see Figure 16 for a real-size representation of these units. The small bars on the upper right in Figure 16 are of 1-cm length, subdivided in intervals of 1 pt, plotted with minimal and with recommended line weights – only to illustrate the resolution capacity of the human eye and the limitation of this printed medium. Note that there is a difference between printer's and computer units. For desk-top publishing, the following relations are applied:

$$1 \text{ printer's pica} = \frac{1}{6} \text{ inch} = 4.22 \text{ mm} \qquad 1 \text{ printer's point} = \frac{1}{12} \text{ pica} = 0.35 \text{ mm}$$

7.1.3   Which typeface and which font to use in graphics?

There are basically three classes of typefaces:

1. the Roman type, characterized by variable line thickness, with the ends of each stroke ornamented by short line segments called *serifs*;

2. non-Roman type, with lines of equal width or weight, without serifs, referred to as *sans serifs* – a type that is quite popular in graphs;

---

[68] An *em* is equal to the point size of the current font, and is the same for all fonts at a given point size: 1 em in a 12 point typeface is 12 points.

3. the default typeface for mathematical symbols and equations used in LaTeX, is known as *Computer Modern*[69] (including symbols like •). CMMI is not the same as TIMES Roman Italic: variables and indices (like $l, 1, l_1$) differ – moreover, sans-serifs are not suitable for math and physics because I and l are too similar.

The principal guideline is to avoid overuse of typeface: there should be no more than 3 to 4 different typefaces in a single research paper.[70] There are lots of other types that are a mix of both basic design shapes. Roman is mostly used for the text body, and sans serif for titles and headers of Sections (see Table 3). Typeface is also used to differentiate between variables, for example in statistics there is a convention to distinguish between *parameter* and *estimate*. The former concept represents the *true* value for the population, the latter is the *estimated* value from the sample: hence σ stands for the standard deviation of a population, and s for the estimate of σ.

When using the Publisher's macros, you should not worry about which typeface to use, and where, because selection of appropriate typeface and font is taken care of in the background. Still, some attention is needed to make the type in the graphics conform with the type used in the text body.

Table 3. Typefaces and fonts used in regular A&A papers. Note that there are only two typefaces, though many fonts.

Element	Typefaces	Font	
Title	HELVETICA	HELVETICA bold	16 pt
Author name	TIMES	TIMES Roman	11 pt
Affiliation	TIMES	TIMES Roman	9 pt
Abstract	TIMES	TIMES Roman	9 pt
Body of text	TIMES	TIMES Roman	10 pt
Footnotes	TIMES	TIMES Roman	9 pt
Sections	TIMES	HELVETICA bold	11 pt
Subsections	TIMES	HELVETICA italic	10 pt
Subsubsections	TIMES	HELVETICA	10 pt
References	TIMES	TIMES Roman	8 pt
Figure and Table captions	TIMES	TIMES Roman	9 pt

7.1.4 Font size

Graphics software provides a font size indicator that allows to select the font size. The font size is the total distance between the highest and lowest point of a set (between top

---

[69]More correctly, CMMI: Computer Modern Math Italic.
[70]The same applies to powerpoint presentations.

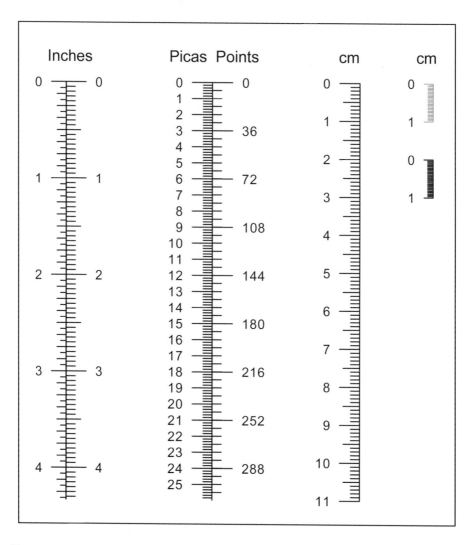

Fig. 16. Typographical units: inches (smallest tick marks $\frac{1}{16}$ [ticks on left side] and $\frac{1}{10}$ [ticks on right side], picas, points and cm. The bars on the upper right, subdivided in intervals of 1 pt (28.34 cm^{-1}), are meant to demonstrate the resolving capacity of the human eye and of this printed medium: the upper sample is drawn at the minimal line width, the lower sample has recommended line width (also called line weight) of 0.3 mm.

of character t and bottom of letter y), and is expressed in points. The width of the letter P in "Points" in Figure 16 is set in 12 pt ARIAL, but is only .75 the height.

Though most books and reports are set in 10–12 pt, you must be extra careful with text inside graphics, because graphs are quite often reproduced at a reduced scale. The smallest type to read in comfort is 8–9 pt, hence text in graphics should be set in 9 pt divided by the reduction factor (the ratio of original linear dimension to printed dimension). This guideline was not followed in Figures 9 and 14, simply because these graphs are included as examples only. The effect of different reduction factors is also noticeable in Figure 9 of Paper III.

A&A 500, 491–492 (2009)
DOI: 10.1051/0004-6361/200912200
© ESO 2009

Astronomy
&
Astrophysics

Special issue

COMMENTARY ON: BERTIN E. AND ARNOUTS S., 1996, A&AS, 117, 393

Modern astronomical surveys and the development of informatics

S. N. Shore

Dipartimento di Fisica "Enrico Fermi", Università di Pisa and INFN-Sezione di Pisa, largo B. Pontecorvo 3, Pisa 56127, Italy
e-mail: shore@df.unipi.it

Fig. 17. Example of A&A typefaces and fonts.

As mentioned before, the Publisher's macro takes care of typeface and font, and prevents you from inadvertent changes of typeface in the text. But font size can be overruled (for example, by decreasing the font size in Figures and in Tables, using `footnotesize` or `scriptsize`. It is good to always remain vigilant to the reduction ratio, so that your textual elements are not reduced to a size that is illegible.

Figure 17 illustrates some typefaces and fonts used on the titlepage of the paper by Shore (2009) in the A&A anniversary issue.[71] The journal name, the title of the paper, and the Special issue stamp, are in the sans serif typeface. All other typefaces are serifs, in different sizes and styles (also small capitals). The same applies to regular papers, where also Abstract and all Section headers are in sans serif, whereas the body of the paper is in serif, see Table 3 that lists the typefaces and fonts used in regular A&A papers.

7.2 Graph title

Graphs may or may not have a title, and quite often space can be saved by using the caption in place of a title. Avoid using the graph name (or the filename together with its extension) that you use internally (*i.e.*, the computer-generated codenames that come with the graph when it is printed to a file). The graph title should never mix with the data and the labels. A larger font may be selected for the graph title, even a different typeface. Figure 12, for example, originally had a title *"Daily sunspot area averaged over individual solar rotations"*, but to avoid visual clutter, the title was removed, and inserted in the caption instead.

---

[71]The pagewidth reproduction factor for this book is $\frac{A\&A}{this\ book} \sim 1.5$.

7.3   Axes

An axis is a base line with respect to which data are plotted or measured. The axis line should stand out, without dominating the graph with excessive line width or improper color. The most frequent mistake is to draw the axis line too thin, *i.e.*, using the minimum line width provided by the graphics computer package (typically 0.01 cm or 0.3 pt). Considering that the human eye can resolve 10 dots mm^{-1} (see also Sect. 8), a 0.01-cm line width is just on the limit – especially when considering that most graphs are scaled down in the printing process, and that print resolution also depends on the quality of the paper medium. To illustrate this point, I refer again to Figure 16 where the three vertical axes have line widths of 0.03 cm (~1 pt). The short segments to the right of the cm-axis have different line widths, and illustrate the problem of using very thin lines (*hairlines* in printers' jargon).

The point where the axes intersect (the origin) often is the point ($x = 0, y = 0$), but it could be any appropriate coordinate value. There is not even a need for the axes to intersect at all.

7.3.1   Axis orientation, and origin

There are no rigid rules about the orientation of axes, but it is quite customary to refer the *x*-axis to the independent variable, with the *y*-axis representing the dependent variable, and to orient the former horizontally. But there may be reasons to deviate from this custom. The *x*-axis is also referred to as *abscissa*, and the *y*-axis is often called the *ordinate*.

In astronomy papers, reverse axes are frequently chosen, in particular when dealing with magnitudes and color indices. Avoid that data points with the lowest ordinate fall right on the *x*-axis: this is not aesthetic, moreover there is a fair chance that symbols placed as such will interfere with the tickmarks (see the inset in Fig. 20 of Paper III). The solution is either to start tickmarking one unit below the datapoint with lowest *y*, or move the axis up or down. An exception is made in Figures 36 and 37, where data points are plotted on the axes, simply because they belong there, and also for the sake of showing the effect.

7.3.2   Axis tickmarks

Tickmarks are short dashes on an axis line to mark the scale along the axis. Together with the units, they help to estimate the numerical values of the plotted quantities, though the main purpose of most graphs is not to read off coordinates, but rather to show relationships between variables. Whether ticks should be on the up or down side of the horizontal axis (or left/right on the vertical) is a preference of taste, see for example Figures 24 and 25. The prime rule is that **tick marks should never interfere with the data** (for an example of such interference, see the leftmost data point in Fig. 42, and also in Fig. 20 of Paper III).

Tickmarks are a problem when they are too close, especially when too many small ticks occur inside a larger tickmark interval. Duplicate axes (*e.g.* a second *x* or *y*-axis at the top or at the opposite side of a graph) mostly have inverted tickmarks at the same locations as the marks on the axis that is duplicated. Except perhaps when showing time-series data, where it can be most useful to duplicate the Julian Date time axis with parallel units of years or decades. Choosing the tick spacing is important: a time axis, for example, should have tick marks at suitable intervals, *i.e.*, 1, 2, 5 or 10 years or longer, whereas tick intervals of 2.5 or 3 years are not so comfortable to handle. Similar attention is needed when tickmarking hours: you can either use decimal fractions of an hour, or sexagesimal multiples (10, 15, 20) of a minute.

Axes without tickmarks or without units should be avoided, except when the omission is done on purpose, as in Figure 18, which is a schematic diagram illustrating the difference between accuracy and precision discussed in Paper I. The data used for the histogram are the early estimates of the velocity of light[72] shown below the dashed line in Figure 3 of Paper I, thus limited to only one lobe of the bimodal distribution. Note that the bin size has been chosen so as to give a "good-looking" bell-shaped curve, but that the histogram (for whatever bin size that is applied) never approximates a true normal distribution.[73] The dashed line in Figure 18 corresponds to $215\,000 \pm 1\,800$ km s^{-1} resulting from a Gaussian fit to the data. The horizontal arrow bar labeled "precision" indicates the information used for inferring the precision, the other arrow bar illustrates how far accuracy and precision wander apart. Blind statistics thus brings forth an amazing precision of less than 1%, whereas the accuracy of this result is not much better than 30%. Using the remaining data (*i.e.*, same experimental results but with a different value for the astronomical unit), however, yields $314\,000 \pm 4\,000$ km s^{-1}, thus a seemingly higher accuracy but a much lower precision. Figure 30 in Paper III is another example of axes without tickmarks and units.

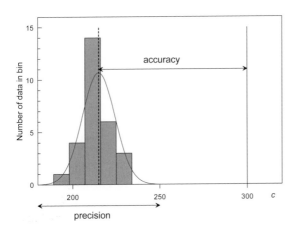

Fig. 18. Difference between accuracy and precision. The bell curve is a Gaussian fit (to a non-normally distributed sample). The solid vertical line corresponds to the value of the natural constant *c*, the dashed vertical line corresponds to $215\,000$ km s^{-1} resulting from the Gaussian fit. The unit on the *x*-axis is 1000 km s^{-1}, tick marks and labels on duplicate axes were omitted on purpose.

---

[72] Allegedly attributed to Ole Rømer, see Paper I.

[73] That is, with roughly 95% of the sample contained in the interval *mean* $\pm\ 1.96\ s$, where *s* is the sample standard deviation.

7.3.3   Axis title and units

The axis title must be in label size or larger, and must be accurate and correct. I have recently coauthored a paper that shows a color-magnitude diagram of an open cluster, featuring isochrones[74] and observational $B - V$ color indices.[75] Evolution tracks and isochrones are theoretical concepts which are, by definition, free of interstellar reddening, whereas color indices of stars are mostly reddened, so when overplotting an observational Hertzsprung–Russell diagram with theoretical curves, the reader must be clearly informed what is exactly plotted, so in this case it can only be the so-called dereddened color index $(B - V)_0$. Unfortunately, I could not convince the leading authors. Surprisingly, the literature on stellar evolution is packed with comparisons of isochrones and evolutionary paths with observational data, and this error is repeated all over: the argument for not correcting is that the meaning is clear from the context. As mentioned in Paper I, **scientific communication should be context-independent**, especially when it is so easily done right.

The axis label should indicate the plotted variable and the units in which it is expressed. The horizontal axis has the label horizontally, for the vertical axis there is the freedom to choose between parallel or perpendicular to the axis. The decision depends on the global aesthetics, and the economy of the page: long labels are preferably aligned parallel to the vertical axis[76], short labels can be put parallel to the horizontal axis.

Data units are sometimes not given at all (as along the abscissa in Fig. 18), or are plainly wrong, especially when units are truncated (*e.g.* JD 4606 for 245 4606) without further explanation. Errors are often made with *multipliers*, for example the x-axis unit in Figure 18 is 1000 km s^{-1}, and since it is awkward to use "kilo-km", the appropriate unit is 10^6 m s^{-1}, which would render the right side of the x-axis somewhat cluttered. One way out is to use the multiplier $\times$ 1000 km s^{-1}, but it often happens that in such a case /1000 km s^{-1} is written instead. Note that some variables are dimensionless, for example phase (ratio of two variables with the same unit) in a cyclic phenomenon.

Breaking axes should be done only when no other solution is possible, because the viewer cannot mentally evaluate the amount of the variable that is jumped, and will often subconsciously perceive the break as a small gap in quantity, see also the discussion of Figure 41. Figure 22 in Paper III is an example where the upper x axis includes a change in scale, whereas the lower x axis keeps one single uniform scale. Such a situation can lead to confusion, hence such difficulties should be mentioned in the caption.

Beware the abbreviations for units, and only use the standard SI nomenclature.[77] Table 4 lists the SI base units (that are dimensionally independent) and Table 5 gives the most common derived units used in astronomy. Note that prefix (hecto, mega,...) and unit form a single symbol that may be raised to a power without adding parentheses. The following list is a set of items related to units:

---

[74]In the context of stellar evolution, isochrones are lines of constant age, across a range of masses, in the Hertzsprung–Russell diagram.

[75]$B - V$ is a color index of the Johnson UBV system.

[76]That is, at an angle of 90° counterclockwise from the x-axis.

[77]International standard ISO 80000 is the successor of ISO 31, a style guide for the use of physical quantities, units of measurement and formulae in scientific papers. SI: International System of Units.

Table 4. The 7 SI base units.

Quantity	Unit Name	Symbol
length	meter	m
mass	kilogram	kg
time	second	s
thermodynamic temperature	kelvin	K
luminous intensity	candela	cd
electric current	ampere	A
amount of substance	mole	mol

- wavelength is no longer expressed in Ångstrom (Å) or micron, but in nano- and micrometer (nm, respectively, μm);

- wavelength λ in meter[78] and frequency ν in Hz are linked: $\lambda = 2.99792458\ 10^8/\nu$;

- when expressing magnitudes in a specific wavelength band or photometric system, it is not necessary to add "mag" or superscript m, since the expression $V = 9.0$ or $B - V = 0.25$, implies that the unit is magnitude;

- some symbols for non-SI units may lead to confusion: for example, straight quotes are used for inch and foot, but also for arcminutes and arcseconds (as well as for beginning and end quotes in a text);

- when using Julian Date, specify if the heliocentric correction has been applied (see the comment on page 153);

- when truncating Julian Date JD 245 5898.5 to 5898.5, the column header should list JD−245 0000 and not JD+245 0000;

- avoid Modified Julian Date (MJD), which can lead to 0.5-day confusion;

- when using Universal Time (UT), specify if it is UT1 or UTC[79] (see Fig. 21);

- special attention is needed for time units when using time format yyy/mm/dd and yyy/dd/mm: 2010/01/10 and 2010/10/01 are different dates;

- the same for decimal hours: hhmmss and hh.xxxx can lead to confusion;

- other formats for time and date should be carefully examined, see for example the case given in the footnote on page 135;

- another confusion with time is that some time signals are recorded in seconds, and others in decimals of a day, with a multiplier of 100 000: since a day has 86 400 sidereal seconds, one second is approximately $0\overset{d}{.}00001$, and the close match with $\frac{1}{100\,000}$ can be easily overlooked, and can thus easily lead to almost undetectable errors in frequency spectra of time series;

---

[78]The US spelling of *metre* is meter.
[79]UTC is a time standard based on International Atomic Time (TAI) with leap seconds added at irregular intervals to compensate for the Earth's slowing rotation. UT1 is defined as mean solar time.

- confusion between e, the base of natural logarithms, and e, the elementary charge;

- a space must be inserted between the numeric value and its following unit symbol,[80] and also between adjacent unit symbols (*e.g.* N m) – except when used in combinations like 1-m telescope, 8-pt characters, etc.;

- avoid mixing decimal separator and decimal comma (*e.g.*, 9.5 and 9,5) in text and graphics;

- when a number is less than unity, include 0 in front of the decimal separator;

- trailing zeroes are only allowed if they are significant, see Figure 52 that reproduces a Table with one and two decimals for the same variable, and where trailing zeroes were omitted;

- verify the correct use of upper case and lower case units: K is Kelvin, k is kilo;[81]

- symbols of units are not abbreviations, and therefore must never be pluralized (*i.e.*, write km, not kms), nor ended with a period;

- only one *solidus* (/) is allowed in compound units, unless parentheses are used, for example $N/(m/s^2)$;

- the prefix and the unit form a whole (like μm), as such an exponent refers to all elements of this combination, see the (inverse) wavelength unit on the x-axis in Figure 19.

It is not forbidden to use different units for the same variable, for example in the case of multiple axes: multiple abscissae or ordinates (like time units, magnitude versus flux, or wavelength versus frequency) are very useful – duplicate axes are even very functional for the purpose of framing the graph. Figure 25 shows an x-axis with a somewhat confusing unit: the label indicates "Years (–BC/AD)", the time unit being year running from 4 000 BC to 2 000 AD, though the negative numbers are self-explanatory and could do without further specification.

7.3.4   Axis type

Not only the design type of a graph, as described in previous paragraphs, is to be considered, but also the most suitable axis format should be chosen. The most common axes are Cartesian, but some data – especially those of a cyclic character – deserve polar coordinates instead. In such graphs, the radial coordinate is defined in $R \in [0, \infty]$, and the angular coordinate in $R \in [0°, 360°]$ or $[0, 2\pi]$. Figure 40 shows a vintage polar coordinate graph with grid.

Phase diagrams, in principle, have x-coordinates in $R \in [0, 1]$, where the reader must mentally link the right and left edges. As mentioned already, a redundancy is quite often constructed by showing, for example, $x \in [0, 1.4]$ or $[-0.2, 1.2]$, a redundancy created to help the eye.

---

[80] A space character always stands for multiplication; preferably use unbreakable space in LaTeX.
[81] That is, the prefix, not the symbol for kilogram.

Table 5. Some supplementary and derived SI units used in astronomy.

Quantity	Unit Name	Symbol	Equivalent
velocity			$m\,s^{-1}$
acceleration			$m\,s^{-2}$
angular velocity			$rad\,s^{-1}$
angular acceleration			$rad\,s^{-2}$
frequency	hertz	Hz	s^{-1}
force	newton	N	$kg\,m\,s^{-2}$
pressure	pascal	Pa	$N\,m^{-2}$
work, energy	joule	J	$N\,m$ or $kg\,m^2\,s^{-2}$
momentum			$N\,s$ or $kg\,m\,s^{-1}$
power	watt	W	$J\,s^{-1}$
electric charge	coulomb	C	$A\,s$
luminous flux	lumen	lm	cd sr
illuminance	lux	lx	$lm\,m^{-2}$
time	Julian year	y^1	365.25 d
	second	s	
	minute	m	
	hour	h	
	day	d	86 400 s
arc	degree	°	
	arcminute	′	
	arcsecond	″	
	milliarcsecond	mas	
plane angle	radian	rad	
solid angle	steradian	sr	
wavelength	nanometer	nm	$10^{-9}\,m$
	micrometer	μm	$10^{-6}\,m$
energy	electron volt	eV	$1.60\,2177\;10^{-18}\,J$
distance	astronomical unit	AU	$149\,598\;10^6\,m$
	parsec	pc	$30.857\;10^{15}\,m$
mass	solar mass	$M_\odot$	$1.9891\;10^{30}\,kg$
flux density	jansky	Jy	$10^{-26}\,W\,m^{-2}\,Hz^{-1}$

[1] Unfortunately there is no official symbol for the year, though "y" can be used.

Not all parameters or observables can be ordered along a numerical axis: some follow rank ordering – a so-called *ordinal scale* – that assigns values to objects based on their ranking. MK spectral types, for example, are mostly represented with a decimal subdivision, but especially in the "wings" of the spectral sequence (O 5–B 0 and K 5–M 0), spectral types are to be seen as labels, and not necessarily as numerals (Wing 2011).

7.3.5   Axis labels

A common problem with Figures is that the axis labels are often too small, illegible, in-consistent, or ambiguous. Figure 25 is an example with appropriate (but different) axis label and axis title font sizes. Incomplete labels are not uncommon, see Figure 26 where labels have been partially clipped in the reproduction process. Together with the fact that the overall image resolution is quite often on the low side, small labels – at reproduction ratios of 0.7 or 0.8 – will complicate the understanding of the graph.[82] A rule of thumb is to aim at 12 points for label size, as was applied in Figure 16. Another problem is miss-ing labels: the *Astronomical Society of the Pacific* (ASP), in its *Overview of Publishing Process*[83] explicitly asks authors to check for *"labels missing around figures"*.

Another common error is forgetting axis labels altogether – though omission can also be done on purpose. Typical examples are the histograms that are inspected during image processing: the image-analysis software will mostly produce histograms with unlabeled axes without tickmarks, because these diagrams are constructed for qualitative inspection only, and are not meant for reproduction. One such example is Figure 28, where the y axis is unlabeled and without scale (assuming that the reader knows that the ordinate is normalized). The same with the x axis, except that it is often indicated that the pixel values run from 0 to 255 from left (dark) to right (bright). An additional scaled and labeled x axis was added for clarity.

7.3.6   Axis scale

The extent of a graph (along all axes) should cover a slightly larger range than the data. The scale can be linear, or non-linear, or the scale can appear linear when the plotted vari-able is not (for example, stellar magnitude is an implicit non-linear – that is, logarithmic – quantity). A non-linear scale means that data are plotted as a non-linear function of the independent variable. It is important to consider – at a very early stage – whether the natural scale of the data really is the optimal scale for presenting the graph, or whether the argument should be a function of the independent variable.

A logarithmic scale can be helpful when the data cover a large range of values. The base of the logarithm has to be specified, for example ln, $\log_{10}$ or $\log_2$. But the data range is not the only determinant for the choice of a particular non-linear scale: the shape of the function or relation to be plotted is crucial, as the subsequent discussion shows.

Stebbins & Whitford (1945) determined the mean colors of unreddened main-sequen-ce stars grouped by spectral type, computed black-body colors for six wavelengths from Planck's formula,[84] and plotted the color indices as a function of $1/\lambda$. Figure 19 shows a reconstruction for B0, A0, G0 and M1 main-sequence stars.[85] The graph demonstrates at one glance that

---

[82] The effect of reproduction ratio on label size is clearly visible in Figure 9 of Paper III.

[83] http://www.aspbooks.org/publishing_overview/

[84] $E(\lambda, T) = \frac{2hc^2}{\lambda^5} \frac{1}{e^{\frac{hc}{\lambda kT}} - 1}$, where h is the Planck constant, c the speed of light, T the temperature of the black body and λ the wavelength.

[85] The calculated indices are not shown, but they agree fairly well with the observations (most of the points fall within the area occupied by the plot symbol, except for the leftmost point for spectral type A0).

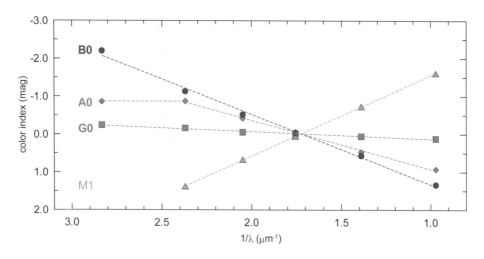

Fig. 19. Example of non-linear axis scales. Observed color index (logarithm of the ratio of two stellar irradiances) for main-sequence stars of different spectral type, as a function of $1/\lambda$. The dashed lines are linear fits that visualize black-body spectral energy distributions (for spectral type A0 the fit is constrained to $1/\lambda < 2.37$). Stebbins & Whitford (1945) list no data for M1 beyond 422 nm. Source: Sterken *et al.* (2011).

1. color index (logarithm of the ratio of stellar flux in two adjacent wavelength bands) is almost linear in $\frac{1}{\lambda}$;

2. some spectra show strong deviations, notably in the photometric band that contains the Balmer jump, and

3. the color indices have a common zero point at $\frac{1}{\lambda} = 1.75$ (570 nm).

This graph shows the power of selecting proper axis scales to render a complicated physical law as linear segments. The graph also displays the zero point of the color index scale.

    There are no rigid rules about which type of axis scale to use, but if the data seem to follow a specific function, using a scale of the inverse function usually renders linearity. This approach is called *linearization*: for example, if the data follow an exponential law $y = ae^{bx}$, plotting along the inverse function $y^{-1} = \ln(a) + bx$ renders the plot linear. Logarithmic scales are the most popular non-linear scales because they are perfect for handling exponential and power-law information. Take, for example, Figures 7 and 8 of Paper III, which show the same bibliometric data on a linear and on a logarithmic scale. One of the datasets (case F) has a citation accretion curve that is strictly linear (on a logarithmic scale), which means that there is an *exponential growth* of this scientist's body of citations. Non-linear scales have a disadvantage, though: they are not as easily readable as linear scales (the logarithmic scale is notable for perceptive exaggeration of small numbers), and tickmarking is not always obvious.[86]

Figure 31 illustrates the use of linear and non-linear scales in the same graph. The x axis is expressed in linear units (arbitrary laboratory scale), the duplicate axis shows the corresponding physical units (nm) obtained through a third-degree polynomial fit of the data given in the original paper. Because the transformation is approximate, the tick marks are shown as $\diamond$ symbols. To avoid visual clutter, not all wavelength marks have been labeled; this example moreover illustrates the mixing of numerical and symbolic axis labels.

One aspect that relates to units is autoscaling, which can lead to a distorted image of the data. Autoscaling accommodates all data in the plot database, including outliers (errors of measurement, mistyped entries, flares, eclipses, etc.) and, as such, compresses the regular data to a scale that is not always appropriate for visual analysis. This leads to variable and inconsistent scales on the axes of different graphs belonging to one piece of work, rendering any comparison impossible.

7.3.7   Axes and framing

Two axes will form a so-called *half frame*, see Figure 43. Sometimes a full frame is formed by four functional axes, for example in Figures 39 and 42, but a frame is sometimes drawn for purely decorative reasons, see Figures 5 and 17.

7.3.8   Background colors and lighting conditions

The background color is an important part of the graph because color perception is affected by surrounding hues. Quite a large number of graphs are made with black or colored foreground symbols on a white background, simply because we print on white paper. The background color is an element to be decided at the design phase. Figure 20 illustrates the disturbing influence of some background colors: color perception changes when colors are viewed on differently-colored backgrounds, not to speak of the problems that color-vision deficient viewers may have, see Section 9.4.5.

Background colors and gray shading are always tricky, especially when printing in black and white. Figure 31, for example, shows a graph with intentionally applied gray shading for the mere purpose of making the yellow spectral line visible. This solution is not optimal for printing on white paper (yellow on white is almost invisible), but the yellow contrasts well on the gray background. Figure 21 is a beautiful example of a creative composition, here slightly adapted to avoid interference with the colored background.[87]

Another important element is the illumination conditions when viewing color combinations. In subdued light, the Purkinje effect (the tendency for the human eye to detect blue objects more readily than red under low light conditions) effectively modifies color content of an image, whereas a bright source of illumination may render colors very differently, as is illustrated in Figure 48.

---

[86]Standard tick values between major ticks on a $\log_{10}$ axis are 2, 3, ..., 9, and divide the interval in 9 parts.
[87]The original graph is available at http://jjy.nict.go.jp/index-e.html

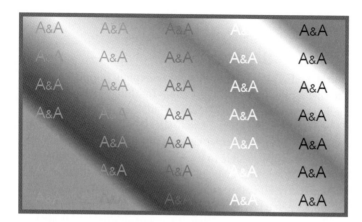

Fig. 20. Example of background contrast problems using, from left to right, *RGB*, white, and black labels. The typeface is HELVETICA, the font size 14 pt. The reduction factor is 0.365, what results in 8.6-pt printed characters.

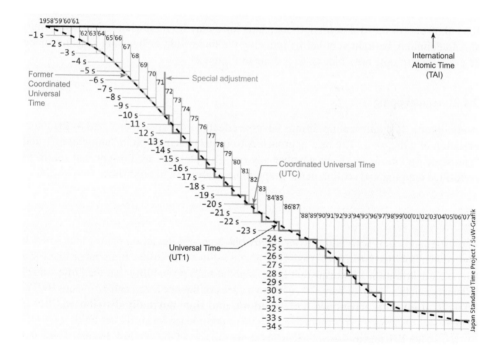

Fig. 21. Difference between uniform International Atomic Time (TAI) and Coordinated Universal Time (UTC), respectively UT1. The dashed curve 1958–1972 is the former Coordinated Universal Time. On January 1, 1972 the difference between TAI and UTC was set to 10 seconds. This graph is a timeline. Adapted from *Sterne und Weltraum*, October 2009, with permission (the only adaptation is a slightly different color rendering for enhancing contrast).

7.4 Plotting

7.4.1 Plot symbols and lines

Plot symbols should be neutral, and signs with an offensive or explicit political connotation should be avoided. The size of a plot symbol must allow for differentiation between several plotted variables, at the same time the size should be small enough to avoid overlapping symbols (see the overlapping data for the small slices in Fig. 8). Differently shaped symbols can be used to distinguish different variables in the same graph. When colors are used for coding, it is still indispensable to use shape as a redundant element of information to identify the variables, because color information is often lost in printing, and because not everyone can perceive all colors optimally (see Sect. 9.4.5).

Symbols can be omitted, and replaced by connecting lines (see also Sect. 11.5) when the symbol density is too high (see Fig. 41). Lines can either be full lines, or dashed, or dash-dotted. The most important point about lines is their width, especially when graphs are reduced in reproduction. Many graphs in this chapter have lines with native widths of 0.03–0.05 cm (or 0.8–1.4 points), though axis lines may be somewhat lighter. Grid lines should definitely be lighter, either by reducing the line width, or by using a lighter shade of gray (for example 60% black), as is done in Figure 21.

7.4.2 Autoclipping

Autoclipping, like autoscaling, brings adverse effects to the appearance and to the interpretation of a diagram. The best approach is to inspect each graph in "autoclip off" and "autoscale on" mode to check whether any out-of-scale data are present, but return to controlled clipping and scaling in the graph to be analyzed and published.

7.4.3 Error bars

The usual approach is to add plus or *minus* one standard deviation (or σ) bars to each plotted symbol, but 2-σ and 3-σ bars are also used in some cases. Remember that for a Gaussian distribution 1σ means that there is about 68% probability that the "true value" falls between the 1σ error bars, 2σ refers to a 95% confidence level, and 3σ means 99.7% – that is, **if all data errors are truly random, and thus normally distributed**. But if there is any systematic error involved, then these confidence levels do not hold.

It may happen that the error bars are of about the size of the symbol. In such a case the solution is to omit error bars altogether, and somehow quantify the data errors in the text or in the caption. As an alternative, you may display one representative error bar (without data point) at a location that is sufficiently off the locus of the data.

7.5 Graphics and texts as image components

Textual components in graphs, such as legends or keys, are often placed inside a legend box (see, for example, Fig. 39). When adding text, you must decide on the font type, on the font size, and also on the font style. It is very important to make text inside graphs consistent with text inside the body of the manuscript, especially with relation to mathematical symbols and units.

7.5.1   Legend or key, and labels

The legend (or key) explains the meaning of the different symbols or colors in a graph. If the graph contains lines, they may also be referred to in the legend. In Figure 19, I used direct labeling of points, so the legend was omitted on purpose, because it interfered with the data symbols. The problem of low contrast of yellow (typically selected for representing stars of class G) was circumvented by selecting brown.

It is best to avoid "ALL CAPS" fonts, because they have all nearly the same height and width, and are sometimes difficult to read. Emotionally, such fonts convey the impression of exclamation marks – which is one of the reasons why I use "SMALL CAPS" for some of the most relevant citations in Papers I–III, a choice that permits omission of leading and trailing quotation marks. Care should be taken that the legend really stands out on the background of the graph. Especially in the case of dark backgrounds (photographic negatives), one must make sure that all annotations stand out in proper contrast. Such textual elements (including arrows) should never interfere with the data elements (such as stars), and should never change the data content of the graphics. For an illustration, see Figure 38, where the labels for the 10–50% densities are dark on a relatively bright background, and the reverse for labels covering 50–100%. A similar change in label color is displayed in Figure 34, and in its color-transformed version in Figure 35.

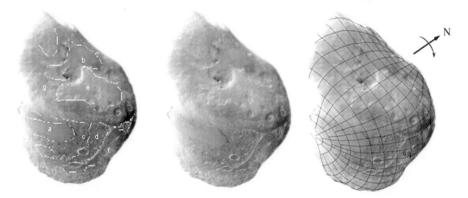

Fig. 22. Example of optimal graphical elements: Comet 9P/Tempel 1, as seen by the Deep Impact cameras. *Left:* with boundaries of layers marked, and identified by small letters. *Center:* without markings, for clear viewing of the surface features. *Right:* with the equatorial coordinate system superposed for easy reference in the text (after Belton *et al.* 2007, with permission).

Labels inside graphs are often added automatically, for example near contour lines. The software gives the user the possibility to define label placement, typically by specifying amounts above, below, left or right from the plotted point. But in some cases such automatic labeling leads to overlapping of labels and data information, and such graphs need more attention, even manual adjustment. Take for example Figure 33, where the wavelength labels along the horseshoe curve have been placed with three different settings – even including reversing font color – to avoid overlap of curves and labels. Labeling of all points in Figure 46 would lead to overlap of labels in the upper panel, and to overcrowding in the lower panel; therefore the two planets between Mercury and Mars were left unlabeled.

7.5.2   Scale factors and orientation

In astronomy (and in all geosciences) graphics frequently need two additional elements: the scale factor, and the orientation. As such, Figure 22 is a nice example of nearly perfect graphical communication: the image is of high resolution, and it provides an orientation arrow as well as a spin-direction arrow. The left image highlights the cometary surface features selected by the authors, the right image explicates the coordinate system, and the middle panel shows the natural and original view, so that the reader cannot be misled by the author's interpretation. For optimizing print contrast, I inverted the grayscale and, therefore, the caption mentions "after Belton *et al.* (2007)".

   I obtained permission to reproduce from first author Belton, but not without first solving a most confusing issue: downloading Belton *et al.* (2007) through ADS delivers a PDF file carrying the bibliographic information "*Icarus* 191 (2007), 573–585" with the small print in the footnote mentioning *"This article originally appeared in Icarus Vol. 187/1, March 2007, pp. 332–344. Those citing this article should use the original publication details."* I first had overlooked this warning, fortunately the author informed me that my reference was not the correct one. We thus have here a very bad example of bibliographic confusion – with neither the author nor the reader being aware of it, causing needless email exchanges between the author and the user.

7.5.3   Insets

An inset graph is a graph within a graph, a smaller-scale picture inserted within the bounds of a larger one. Figure 3 illustrates the application of an inset: by using space without information content inside the larger graph, a vintage diagram is included at a smaller scale – though with preservation of resolution.

   If an inset refers to a result or an argument from within the same paper, all its elements (labels, legends, . . .) should be of appropriate font and size. If the inset is a reproduction of a graph published elsewhere, the size can be downscaled regardless of font size, provided the full reference to the original source is given (as is done in Fig. 3).

   Figure 23 is a good example of proper combination of data and inset graphics: the dynamical orbit of the young massive binary system θ^1 Ori C is illustrated (Fig. 8 of Kraus *et al.* 2009). The available astrometric data and their error bars are shown, as well as $O - C$ vectors[88] (arrows). The original Figure shows multiple orbit solutions. The nine insets (interferometric images) illustrate the data, and contribute to a most decorative composition.

---

[88]$O - C$ stands for Observed *minus* Calculated.

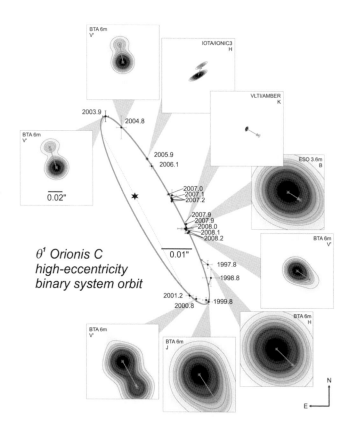

Fig. 23. Example of proper combination of data and decorative graphics: the dynamical orbit of a young massive binary system. Based on Kraus *et al.* (2009).

7.5.4   Grid lines

Grid lines are remnants from the gridded graph paper used in pre-computer days (see Fig. 40), and these grids were quite often printed with a special green or light blue ink that would not reproduce on a XEROX copy or a print. Grid lines are now a standard and automatic aspect in business graphics, but in scientific graphs of high data density, grids often form visual ballast that serves no purpose, simply because the reader will not try to derive precise quantitative information from graphs. The more so because the younger generation has never been accustomed to use grid lines, so their visual-mental perception system is not well developed in suppressing the grid lines and seeing only the data. In short, grid lines – especially heavy grids – unnecessarily clutter the diagram and get in the way of seeing the data: they are to be considered as chartjunk, unless they are faint, really helpful for reading off numerical data values, or for enhancing visualization, see Figure 21 and also Section 4.4.

Figure 24 shows Figure 1 of Holweger & Oertel (1971) in its original form, displaying 15 redundant grid lines, together with the same graph with the grids removed. Non-cartesian plots (polar and ternary diagrams) may, nevertheless, profit from light-shaded gridding because their axes are less commonly used, and the reader needs more effort to

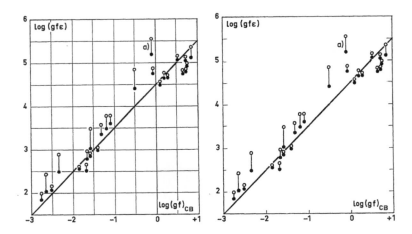

Fig. 24. Example of graphics with and without grids. Based on Figure 1 of Holweger & Oertel (1971).

assess the data. As mentioned already, 3-D surfaces are often overplotted with a mesh grid, though in Figure 9 the grid lines were omitted on purpose, with the result that the reader cannot tell whether a peak exceeds ordinate value 100 – the more so because the color code is not included.

These principles equally apply to images, graphs and tabular material. But functional grid lines can be useful. Figure 25 is taken from a paper by Usoskin *et al.* (2007), and illustrates another example of a very good diagram: sunspot activity through the Holocene is shown,[89] and the two horizontal lines represent crucial data levels, not just grid lines. The diagram cannot fail to direct the eye to a most important fact: the present-day grand maximum in the decadal mean sunspot index, the highest level in more than ten millennia. Beware, though, that this graph is based on sunspot data supplemented by proxy data records. This graph has a simple composition, nevertheless the scientific message stands out very clearly.

7.5.5 Decoration

In genuine data graphs there is not much reason to add decorations, but it is acceptable, as long as decorative elements do not overload the graph, and as long as they do not change any quantitative aspect of the data. An example of inclusion of a decorative banner is shown in Figure 5.

---

[89]Only the lower part of the original is reproduced. This diagram figured on the cover page of the first issue of A&A Vol. 471.

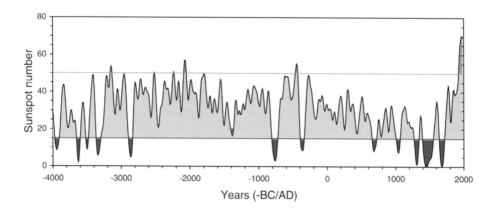

Fig. 25. Reconstruction of decadal average sunspot numbers through part of the Holocene. The *x*-axis label is explained on page 106. Source: Usoskin *et al.* (2007).

7.6 Aspects of composition

7.6.1 Galleries

Galleries are compositions – often ordered in an array – of several smaller images that deal with the same aspect of the data (for example, a set of finding charts or spectra). If possible, all panels in a gallery should have the same size, and the same amounts of white space between horizontal and vertical boxes, so that all elements of the gallery should render in a neatly tiled pattern. A simple case of gallery arrangement is Figure 22, which can be made as a single image file, or can be combined by compiling three separate image files side by side in the manuscript's LaTeX file. When doing so, one must make sure that excessive white spaces are trimmed off each image, otherwise these spaces will add up and create too large gaps between the image components. Such gaps are not only inaesthetic, they also render images too small in size. Images should thus be cropped, and preferably be oriented in the same way (portrait or landscape) as intended for print (for example, Figs. 27 and 43).

When building galleries of images (especially negative finding charts), the matrix of gray squares on a white background can produce the so-called *Hermann grid illusion*,[90] with dark areas appearing at the intersections of the white spacer lanes. The gallery in Figure 27 will – under specific ambient light – invoke such an illusion when viewed on a computer monitor.

---

[90]Observed by Hermann (1870), in a German translation of John Tyndall's lectures.

7.6.2   Size ratios

There are no rigid rules about the size ratio h/w (relative height and width) of graphics. Some aspect ratios[91] are "hardware-defined", for example visualizations of an imaging detector's field of view, or a map of an elongated celestial object. Scatter and line plots may also have aspect ratios that are conditioned by external factors: color–color diagrams, for example, are usually depicted with the same magnitude scale on both axes. LaTeX, unfortunately, forces (or seduces) the author to use the full page width for graphs. The following aspects should always be checked:

- tall and thin graphs are to be avoided, if possible;
- some designers swear by the *golden section* $\frac{h}{w} = \frac{1}{1.6}$, a so-called pleasing proportion,[92] though there is no good reason to apply this strong belief, except for the fact that such a ratio minimizes wasted space on both sides of illustrations that do not use the full page or column width (unless text can be set adjacent or wrapped around);
- changing the reproduction h/w ratio should be avoided, see the consequences in Figure 26;
- check the reduction ratio: for this book, the ratio is 1:1, hence Figure 16 should have correct metrics;
- take maximum benefit of the page-size ratio. For this book the ratio is $\frac{h}{w} = \frac{1}{1.6}$, for an A&A full page it is $\frac{1}{1.35}$, and for a single column it amounts to half as much.

Fig. 26. Illustration of adverse effect of changing the h/w ratio: dimensions of axis labels and titles have changed along. Note the clipping of the lower part of the *x*-axis labels. *Left*: bitmap image, *Right*: grayscale rendering.

One element to check further is whether the aspect ratio unwittingly influences the reader's perception: high h/w ratios may leave the impression that a given slope or trend is much steeper than it would appear in a graph with a low h/w ratio. This facet is even more important when two adjacent graphs with different h/w ratios are compared.

    Stretching and squashing graphs, when all objects are grouped, will distort all labels, and circular pie charts will become ovals. In single-object graphs, even modest compression will become visible because also the grid lines are affected. Undesired and sometimes undetected squinching is sometimes introduced by the scanning process, and therefore it is required to test your scanner at least once by scanning a page with grids only.

---

[91]The aspect ratio is the ratio of longest to the shortest image dimension.
[92]*i.e.*, movie-screen like.

7.6.3 Figure captions

The caption is a most important textual element – although captions are not composed within the graphical package, but in the manuscript itself. Captions should be brief, at the same time complete. The caption tells the reader how to read the graph, but should not list conclusions and interpretations, nor explain acronyms and abbreviations. A limited amount of data may be included, for example Figure 45 shows a Table embedded in the caption, a composition that allows to save space.

Captions are typeset in the manuscript, and, as such, correct fonts and font sizes are automatically applied. The house style of the journal should be consulted, especially concerning descriptions like *top, right, left, ...*, and whether these indicators should be followed by a colon. The caption is the place to list the permissions to reproduce, or the *"courtesy of ..."* or *"with permission from ..."* statements. If you reproduce existing artwork, the location (museum or archive) and the geometric dimensions of the original have to be given.

When a Figure contains color symbols, it makes no sense to refer in the Figure caption to the colored elements when the publication is in grayscale. As this book is printed in black-and-white, but appears in a full-color digital version, different graphic symbols were chosen to represent colors in such a way that the reference to a symbol allows unambiguous decoding, whatever the color mode of the consulted page.[93] This approach is crucial if you want your graphics to be understood by persons who suffer from color-vision deficiencies, see Section 9.4.5.

7.6.4 Figure numbering and placement

Graphics should be placed immediately after the first mentioning in the text, and may precede the citation only if it is placed on the same page or on the page opposite to the reference. The graphics placement in the present paper deviates somewhat from this rule, because several graphs contain design elements that refer to descriptions in different places, often located a dozen or more pages apart.

If a graph extends over more than one page, the Figure count should be interrupted so that all parts of the Figure refer to the same number, and the caption should clearly indicate this with a statement like "Figure 1, continued".

Avoid large empty spaces above and below the plotted information: such spaces are better used for scaling up the graph to show more detail. These spaces are often produced by an improper bounding box of a POSTSCRIPT Figure, which may even result in partly covering lines of text (or the running title or page number) above the Figure. LATEX cannot handle Figures with text wrapped around, though unaesthetic empty spaces can be avoided by placing the caption adjacent to the Figure, see for example Figure 1 and, especially, Figure 36.

---

[93]The most significant graphs of this paper are also reproduced in quadri on 16 unnumbered pages after page 170.

8   Image quality: high resolution and low resolution

8.1   Image resolution

Every Editor of a proceedings book knows that deficient image resolution is a very common problem. Image resolution refers to the amount of detail, sharpness and clarity of an image, and is quantified in terms of pixels, dots, or lines per unit of length (usually the inch). In a way similar to the confusion between typeface and font, resolution is often referred to with incorrect terminology. Experience shows that the relation between the intrinsic resolution of an image and its print resolution, is often not well understood (*i.e.*, the difference between resolution of the capturing device and the display device, see the discussion of Fig. 27 below). The human eye, for example, has an optical resolution of 0.1 mm or ~ 250 dots inch^{-1}, see the small bars on the upper right of Figure 16 with tick marks at 1-pt interval,[94] illustrating that it is very hard to achieve such a resolution in desktop printing. The terminology on resolution varies according to output device,[95] see Table 6.

Table 6. Image resolution according to output device.

device	typical value	unit	abbreviation
human eye	250	dots inch^{-1}	dpi
screen (websites)	72–100	pixels inch^{-1}	ppi
laserprint	300–600	dots inch^{-1}	dpi
inkjet	up to 1200	dots inch^{-1}	dpi
imagesetter*	2500–5000	dots inch^{-1}	dpi
scanner	at least 150	samples inch^{-1}	spi
halftone	150 and higher	lines inch^{-1}	lpi

*Highest quality printers for professional publishing.

High-resolution graphics was a problem in the past, and computer-generated data have often been printed at a resolving power not better than the line and character spacing of the typewriter/printer (3–4 pt, see Fig. 52 that very much resembles line-printer graphical output resolution of the 1960–70s). This hardware problem is extinct now, nevertheless an amazing number of sloppy-resolution artwork still populates journals and books, mainly because properly-generated computer plots are simply saved to an output file at screen resolution, and printed as such.

Optical resolution is the resolution determined by the optical and mechanical hardware of the capturing device (scanner or camera), and is the true resolution that the device can handle. *Digital resolution*, on the other hand, is an artificial concept referring to resolution arrived at via pixel interpolation, or by means of another transformation. The frequency of lines in a halftone screen, lpi (introduced in Sect. 6.1) also refers to

---

[94]Thus, the average human eye can resolve graphics that are about three times better than this example.
[95]Digital-camera manufacturers sometimes call *capture resolution* the number of pixels of the image sensor.

resolution. That concept, however, is not relevant to most authors, since most of us do not even know the lpi specification of our desktop printers, simply because lpi is not a matter of hardware,[96] but a software issue that is part of the halftone creation process. The three concepts are summarized in the box below.[97]

> **ppi = image resolution: includes only image information**
> **dpi = output resolution: includes output device information**
> **lpi = frequency of lines in a halftone screen or scan**

The pixel is a unit of information, and **has no meaning unless its size is given**, *i.e.*, by specifying the number of pixels per unit of length. Monochrome images require much higher resolution than halftones in order to prevent aliasing (staircase appearance) of diagonal lines. Combinations of shades of gray and text or lines (for example, finder charts with annotations), must reconcile aliasing with file size, and are typically set to 500–900 dpi. Images should also be scaled to the size planned for print, but care should be exercised when resampling: adding pixels by interpolation (resampling up) leads to image degradation, while resampling down leads to deleting information. Resampling also plays a role in image rotation: where rotation over 90° does not change image content (because pixels need not be repositioned over the border between two pixels – that is, when pixels are square), every rotation over other angles will degrade image quality.

Remember that sending 72 dpi to a 600 dpi printer produces output that is no better than that of a 72 dpi printer. Figure 27 reproduces a detail from the book cover illustration: the panels have image resolutions of, respectively, 600, 300, 100 and 72 ppi, and are printed on an approximate size of 51 × 28 mm (2.0 × 1.1 inch). So the print resolutions are, respectively, 300, 150, 50 and 36 dpi. The example shows that the 300 ppi image can be printed at an acceptable resolution if the physical dimension of the image is scaled down by a factor of two (thus a 4× smaller print).

The consistency in resolution inside artwork images must always be considered: if you mount images in one single composition or gallery, you must make sure that the resolutions of all sub-images match – at least when the original images have comparable resolutions. But not only the coherence in resolution in a gallery is important: the viewing resolution of each sub-image must be acceptable. For example, if an online publication offers a 5 × 5-cm 600-dpi graph (mostly JPEG or BMP) with 5 × 10 sub-images, then printing the complete graph at 2× magnification will yield a comfortable 300-dpi print, but a same-size print of one gallery element will result in 60 × 30 dpi, totally inadequate for evaluating or digitizing the data.

---

[96]Except for the tapestry artwork discussed in Section 3.2, with 0.46 lines mm^{-1} in the vertical direction (see page 74), and thus 11.7 lpi.

[97]Dpi is not the size of the dot that the output device produces, but describes the spacing of the grid of dots, also called the *addressability*. Resolution is the smallest feature that can be printed, and is a combination of dot size and addressability. Since dot size is often much smaller than pixel size, dpi must include device information, whereas ppi only has image information.

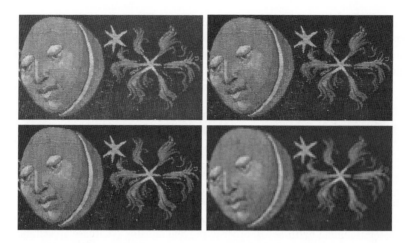

Fig. 27. *Top left:* detail of cover illustration at 600 dpi. *Top right:* same at 100 dpi. *Bottom left:* detail of cover illustration at 300 dpi. *Bottom right:* same at 72 dpi. The lowest-resolution images clearly show ragged outlines and also *pixelation, i.e.,* individual image pixels becoming visible.

8.2   Sampling frequency and scanning

Sampling is the technique of selecting an appropriate sample of a population that is suitable to represent the complete population. In the context of this paper, sampling refers to collecting (image) data at equidistant points. The sampling theorem[98] states that

> IF A FUNCTION $f(t)$ CONTAINS NO FREQUENCIES HIGHER THAN W COUNTS PER SECOND, IT IS COMPLETELY DETERMINED BY GIVING ITS ORDINATES AT A SERIES OF POINTS SPACED $\frac{1}{2}W$ SECONDS APART

(Shannon 1949). In other words, **the sampling frequency must be more than twice the frequency of the original signal**, and consequently one cannot extract from a continuously sampled dataset any frequencies greater than $\frac{1}{2}\Delta$ (where Δ is the sampling interval), because aliases arise due to beating between the sampling frequency and any real frequency present in the data (see also Kurtz 1983 for a discussion related to Fourier analysis in astronomy). Thus, when scanning a photograph intended for 1:1 reproduction, you must sample with twice the frequency of the output linescreen:

$$\mathbf{ppi_{scanning}} = 2 \times \mathbf{lpi_{output\ printing}}$$

When reproduction is on a different scale, the above-mentioned value must be multiplied with the reproduction scale factor (*reduction scale factor* <1< *enlargement scale factor*).

---

[98] Also called the Nyquist–Shannon sampling theorem.

The scanner's resolution qualification is determined by the x-direction sampling rate (the number of sensors in the detector array) and by the y-direction sampling rate (the interval of the stepper motor). Most desktop flatbed scanners have a true hardware (optical) resolution[99] of at least 300×300 ppi. Below are a number of rules-of-thumb for good scanning practice:

- scanning line art (drawings, etc.) and texts must always be done at the output resolution of the final output device, in 2-bit mode, because 8-bit halftones or any other scanner color option will make the image appear smoother – just compare the 8-bit and the 2-bit scan of lineart in Figure 26: the 2-bit result appears crispier;

- consequently, scan at grayscale resolution only when the original is a true grayscale (or color) graphic;

- scan at the target output resolution ($2 \times$ lpi) of the printer, or about half the printer dpi: you can always decrease resolution after scanning, but not the other way around;

- at the same time, make sure that the settings of the video board match the capturing image resolution, $i.e.$, the video settings must be 24-bit and not 16-bit mode;

- images for screen used to be 72 dpi,[100] but nowadays screen resolutions exceed 100 dpi, and thus scanned graphics for the web are increasingly produced at 100 dpi;

- start with high-quality originals and avoid images from books or magazines;

- take care with historical drawings: they may be so faint and specked that no amount of effort can improve them (see Fig. 30, which also shows lines that overlap text);

- scan at the target output size;

- save/compress with minimum loss, better even compress lossless, and

- verify the relation between the scanner output and the image brightness input, using test strips or grayscale wedges.

Figure 28 illustrates a quite instructive example of a scanning experiment. The first version of this book's cover image scan was delivered at 300 dpi resolution, but because the grayscale patch was not very well resolved, I requested a new scan at 600 dpi. Figure 28 shows the normalized histograms for the B channel:[101] whereas doubling the resolution should only change the number of pixels (and the file size) by a factor of four (and hence make the distribution appear smoother), the histograms are markedly different – a fact already obvious from visual inspection since the second image was much darker than the first, see Section 11.4 for more details. Figure 28 is also an example of how straightforwardly decodable color graphics (through its legend), may become less obvious when printed in black-and-white.

---

[99]That is, not the software-enhanced interpolated resolution through the creation of extra pixels in between the scanned ones.

[100]Because the original Macintosh 14″ monitors had a fixed 72 dpi resolution.

[101]See Section 9.4.2 for the meaning of B.

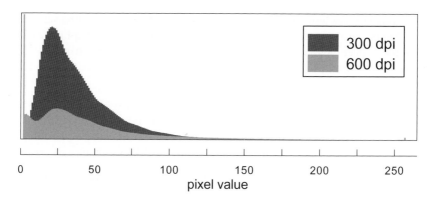

Fig. 28. Normalized histograms of the cover image, based on scans at 300 and 600 dpi resolutions.

9   The complexities of handling color

Most people can see 20 000 to 100 000 colors (in the human visual range, 380–800 nm), and today's computer artists have a palette of over 16 000 000 (*i.e.*, 255^3) colors at their disposal. All these colors come to us as emission spectra of light sources, or reflection[102] (but also transmission and absorption) spectra of materials (paper, glass, plastics, ...). Yet, using more than 2 dozen hues may already lead to bad graphics. Let us first see why we want to use color in our designs, and if all of us are really much helped by doing so.

9.1   Why use color in graphics communication?

Technically spoken, white and gray differ only in level of brightness: gray is not a true color[103] because different grays only differ in luminance, and not in spectral signature. Black, on the other hand, can be described as "zero color" – *i.e.*, no signal at all. Figure 4, for example, uses one color in various shades, and this color view can easily be replaced by a grayscale shading without any loss of information, though with perhaps a less pleasing visual outcome. Some authors are convinced that black-and-white or grayscale graphics can convey any message as well as color graphics can – that is, if the design is well done.

Color simply is an additional – though very powerful – element of information, and if well conceived, color graphs may help your message to better stand out. However, color is expensive, both in printing cost and in memory storage, and haphazard addition of some colors just cannot improve your graphics: **sloppily designed color work is always worse than any black-and-white graphics**. Bad color work can lead to wrong emphasis and misinterpretation, especially for color-blind viewers.

---

[102]See Figure 48 for an illustration of the impact of ambient light on the perceived color of the substrate.

[103]For example, in the four-color-map problem in cartography, gray is not one of the operational colors.

The choice of the color coding is entirely free, although there exist some traditions, for example, using blue for hot and red for cool in astronomical spectra, or the typical color coding of bathymetric and topological maps that use light to dark blue for ocean depth, and green to brown hues for height of landmasses. There are several reasons for using color in graphs:

1. to label items in a graph (color as symbol);

2. to separate distinct data (color as identifier);

3. to emphasize specific parts of a graph (color as organizer);

4. to measure (color as quantitative information);

5. to facilitate teaching (color as didactic tool);

6. to represent the real world (color photography), and

7. to decorate graphics (color as decoration).

There is a basic principle[104] of graphic design:

> **DESIGN IN BLACK AND WHITE**
> **ADD COLOR WHEN THE DESIGN IS COMPLETE**

In other words: use color primarily for emphasis and for conveying additional information, but **do not let color be the only visual differentiation in your graphs**.

9.2   Different colors for everyone

We actually do not "see" colors: a specific visual stimulus activates our color detectors (the photoreceptors called cones) in the retina of the eye, and color is nothing else than a subjective biological sensation provoked by the reaction of the brain to that amount of stimulus. Most human eyes have cone receptors with different sensitivities for short-, medium-, and long-wavelength light, and it is exactly this wonderful physiological property of the eye that is used in all principles of color reproduction in painting, dyeing and printing – even in weaving, see the footnote on James Le Blon on page 75. The color of an object or a light depends on:

1. the spectrum of the light emitted by or reflected from it (see Fig. 48);

2. the ambient light;

3. the observer, and

4. what the observer was previously looking at.

---

[104]This principle is also applied in photography courses and schools: students first follow courses in black-and-white photography before entering the color photography curriculum.

A complete physical description of a color would involve full knowledge of the spectrum of all involved light sources, in addition to the spectral response of the observer's own eyes.

I recently gave a talk at a conference, and that same day, during the closing dinner, a gentle young man came to me and said *"That was a nice talk, with good graphics, but never again use red characters on a green background: I am color blind."* He explained that he could, however, read some of the red characters on green background because I had used large-size shadowed or embossed fonts. Very few people are completely color blind, and most color-vision deficient persons can see colors, though they have difficulty differentiating between particular shades, see Section 9.4.5 for a more profound discussion.

9.3   An early generic mathematical color model

The foundation of the theory of color composition was formulated by Isaac Newton in 1704 in his *"Opticks"*: every color in nature is produced by mixing different colors of the spectrum (of sunlight). Newton divided the circle in seven parts[105] *"proportional to the seven musical Tones or Intervals of the eight Sounds"*,[106] and wrote that the spectrum (from blue to red) appears

> ... TINGED WITH THIS SERIES OF COLOURS, VIOLET, INDICO, BLUE, GREEN, YELLOW, ORANGE, RED, TOGETHER WITH ALL THEIR IMMEDIATE DEGREES IN A CONTINUAL SUCCESSION PERPETUALLY VARYING: SO THAT THERE APPEARED AS MANY DEGREES OF COLOURS, AS THERE WERE SORTS OF RAYS DIFFERING IN REFRANGIBILITY.[107]

Nowhere does Newton put forward that the spectrum contains *only* seven colors (see Fig. 29, which actually is a polar diagram, and also a pie chart) – to the contrary, he explicitly stated that there exists a wide sequence of spectrally pure (*homogeneal*, as he called them) hues that cannot be further split by refraction: monochromatic light as we call it today. This leads to the two basic questions in colorimetry, *viz.*,

1. how many different colors are needed to produce *any* color by mixing, and

2. in what amounts must these spectral hues be mixed?

Of course, artists have long known: mixing blue and yellow yields green, and by mingling red, blue and yellow dyes in different proportions, painters could make almost any color they wanted.

---

[105] Scientific records before 1611 trace four aspects of color diagrams: circles based on seven colors, circles based on four colors, circles organized for a red-yellow-blue mixing system, and historical relationships between pigment primaries and spectral primaries. Parkhurst & Feller (2007) conclude that color organization diagrams are at least as old as the 13th century, with the seven-color circle possibly the oldest found, the four-color next (14th–15th centuries), and the red-yellow-blue circle the most recent (18th century). No graphic records of circular systems earlier than 1611 have yet been identified.

[106] Newton associates color with *intervals of sounds*, but nowadays many sources mention frequencies of *tones*, whereas it is the ratio of *sensations* to which Newton referred.

[107] *Opticks* (1704), Prop. II Theor. II, page 87. Most people can see six colors, and the splitting of purple in violet and indigo is not very widely perceived, as remarked by John Dalton, see page 135. Indigo comes from Latin *indicum*, and refers to India as a supplier of indigo dye.

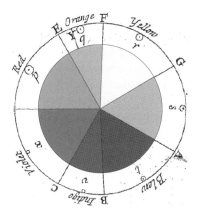

Fig. 29. Newton's color circle (Fig. 11 from *Opticks* 1704, Book I, Part II, Plate III), displaying seven colors selected by Isaac Newton. Probably the first graphical record of the concept of circular color systems. Newton's basic drawing is overplotted here with the 7 *roygbiv* colors: red, orange, yellow, green, blue, indigo and violet.

We owe a lot about color theory to the experiments of James Clerk Maxwell, and especially to his 1855 and 1860 papers. Maxwell's (1855) *Experiments on colour, as perceived by the eye, with remarks on colour-blindness* is even very remarkable as a scientific paper from the viewpoint of the principles explained in Paper I. The title of his paper is succinct, though very complete. The paper has no IMRaD structure whatsoever,[108] but its opening paragraph

THE OBJECT OF THE FOLLOWING COMMUNICATION IS TO DESCRIBE A METHOD BY WHICH EVERY VARIETY OF VISIBLE COLOUR MAY BE EXHIBITED TO THE EYE IN SUCH A FORM AS TO ADMIT OF ACCURATE COMPARISON; TO SHOW HOW EXPERIMENTS SO MADE MAY BE REGISTERED NUMERICALLY; AND TO DEDUCE FROM THESE NUMERICAL RESULTS CERTAIN LAWS OF VISION.

explains what the paper is about, including all elements needed in a structured abstract – objective, methods, results – in fact, a structured abstract *avant la lettre*.[109] Maxwell presents error discussions, describes tossing out outlying data,[110] because two observers "...*namely, Mr R. and Mr S., were not sufficiently critical in their observations to afford any results consistent within 10 divisions*... ", explains boundary applications of the theory (related to the color-defective eye), and the corrections needed for such applications. In this paper he describes his color-mixing experiments that are based on rapidly-rotating color-paper sections,[111] from which he concludes that color perception is not due to the real identity of the colors, but on account of the constitution of the eye of the observer. And that the law of color vision is identical for all "ordinary eyes".

Young (1802) had put forward[112] that "*each sensitive point of the retina contains 3 particles matching the three principal colors, red, yellow, and blue...* " enabling the eye

---

[108] See Paper I, Section 9.

[109] Idiom meaning *before the concept existed* – not to be translated when used in another language.

[110] So-called outliers, see Section 11.4.

[111] *Section* refers to a geometric section, not to a part of a paper.

[112] The Bakerian Lecture, *Scholium*, page 20.

to react to "every possible undulation".[113] Such principal colors are now called *primary colors* or *primaries*,[114] the corresponding receptor types are generically named short, medium and long-wavelength cones, and form what is called a *trichromatic color system*. Maxwell, however, struggled with the fundamental difference between mixing light rays and mixing lights reflected by pigments: mixing dyes[115] makes us see any light that the colorants do not absorb, or do not extract – therefore, color mixing by pigments or by dyes is called a *subtractive process*. In combining light sources, there is no extraction, and all hues combine in what is called an *additive process*.

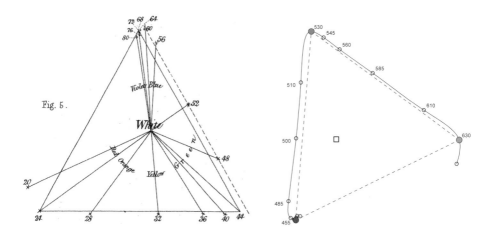

Fig. 30. *Left:* Figure 5 from Maxwell (1860). Channels 24, 44 and 68 correspond, respectively, to **his** selection of red, green and blue primaries. *Right:* rendering of Maxwell's triangular graph, rotated by 145° for comparison with the chromaticity diagram in Figure 33. The central point (□) is the position of "white" as derived from twenty observations. The numbers along the curve are wavelengths of pure spectral hues (in nm), and correspond to Maxwell's channels numbered 20–80.

Maxwell's great discovery was to describe the system that relates any color to the set of primaries, in the so-called *Maxwell triangle*, a ternary diagram where each primary color (red, green and blue) is placed at the vertex of an equilateral triangle, as shown in the left diagram of Figure 30. Inside such a triangle, every color can be made by mixing different amounts of each primary,[116] and white (also called the white point) is obtained by adding maximal amounts of each primary (for an example, see Fig. 48).

Mathematically that was clear, but psychophysically there are two complications that make it impossible to construct *any* color from the three primaries:

---

[113] *Undulation*: hue or wavelength – note that this statement is not strictly correct: not every receptive cell of the retina can distinguish color; some cells (rods) do not have any color sensitivity at all.

[114] Primary colors are colors from which all other colors can be derived by mixing in different amounts.

[115] Dyes are chemicals dissolved in a fluid, pigments are small particles that are not dissolved in liquid.

[116] That is, primaries should be independent: no additive mixture of two primaries should yield the third.

1. the primaries are not (in fact, need not be) exactly known, and

2. primaries are not monochromatic, but have a substantial *bandwidth* that includes neighboring hues (see Fig. 31).

Maxwell's concept is called a *trichromatic color space*. Let me just make a sidestep, and remark that the definition of such a trichromatic color space is very similar to the definition of a photometric system in astronomy, such as the Johnson *UBV* system (that also has a set of three overlapping wide passbands).

Maxwell (1860), on the other hand, is a fully-structured paper with introduction, history, method, description of instrument,[117] method of observation, data reduction, error discussion, results, comparisons, conclusions, and even a postscript note on color-deficient observers (but no Abstract, although the last Section of the Introduction lists the structure of the paper). His paper refers in detail to previous work by Newton, Young, Helmholtz, Fraunhofer and Poggendorff, and sets the stage for his own research: his experiments demonstrate that the chromatic observations leading to color equations can be made with great accuracy, both by normal and by color-deficient eyes. In a closing note, he explains the difference between Newton's theorem on the mixture of colors[118] (Newton's color circle divided into seven parts, Fig. 29), and Maxwell's own model.

For this investigation he does not work with rays reflected from a spinning color wheel, but with a purpose-designed optical device that allows him to combine red,[119] green and blue rays to render shades of white. From numerous measurements of the whites at 16 spectral locations, he constructs color equations that yield white light, and obtains intensities of the three standard colors of *"observer J"* (James himself), which he subsequently marks as R, G and B. Amazingly, he arrives at describing all experiments and their results in words, and in pure black-and-white graphics.

Maxwell's experiments led to the first mathematical (trichromatic) color model. Figure 30 reproduces the results for his own eyes in a ternary plot, where each of his selected primary colors (R for scarlet, G for green and B for blue) is a corner of the triangle.[120] Points outside the triangle result from negative components of one primary; measurements in the blue-violet corner of the triangle are quite difficult, and led to more erratic results. The rightmost panel is exactly the same diagram, but with the location of Maxwell's pure spectral hues connected by a spline curve (see Sect. 11.7). Today we call such a diagram a *chromaticity diagram*, and the 3-dimensional space it describes is called a *color space*. A chromaticity diagram tells us only which mixtures of light will produce a specific color. Figure 30 thus shows James Clerk Maxwell's personal color space, or the area delineated by the triangle of the primaries that he selected. The range of colors that such a model can represent is called the *color gamut*.[121] His 1860 paper also contains the numerical gamut for another person (*"observer K"*), and Maxwell comments on the

---

[117]Maxwell expertly used standard equipment, but also developed his own apparatus.

[118]*In a mixture of primary colours, the quantity and quality of each being given to know the colour of the compound.* Newton, *Opticks*, Book I, Pt. II, Prop. VI.

[119]Maxwell notes *bright scarlet* (vermillion).

[120]Though not immediately obvious, this drawing is also sort of a pie chart.

[121]The word *gamut* originated in music, and nowadays refers to the whole range of pitches an instrument or a voice can reach.

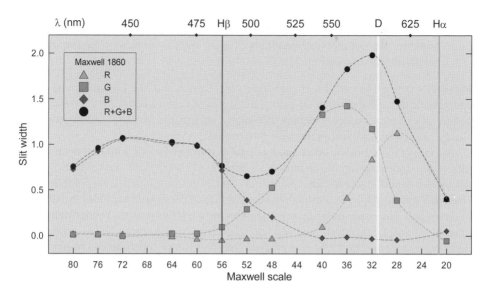

Fig. 31. *RGB* spectral response curves for James Maxwell's own eyes. Dashed lines are spline curve fits. The vertical line near scale value 31 is the doublet of the sodium spectrum near 589 nm (rendered in yellow and thus almost invisible on a grayscale print). Hα and Hβ (Fraunhofer C and F, respectively), are also indicated. The linear scales of the *x* and *y* axes are in laboratory units. The duplicate *x*-axis is expressed in wavelength on an approximate (nonlinear) scale, with small ♦ symbols instead of tick marks.

"blindness" of his very own eyes to parts of the spectrum: *"the feeble intensity of certain kinds of light as seen by the eyes of J ... "*.

Thus, *trichromacy* is a property of ourselves, and the color space as shown in Figure 30 is unique, and James Maxwell's only. Or in modern terms, *device-dependent*: every person has her/his own color gamut, just like every input or output device also has its own gamut,[122] in fact one could say that each device has *its own language for colors*. Color – like beauty – is in the eye of the beholder,[123] and is not a synonym for wavelength.

Figure 31 shows Maxwell's empirical *color-matching functions*, as well as the sum for each retinal receptor type. His general conclusions list that his experiments show that the number of primary elements of color (for the normal eye) is **only** three, and that the *exact* wavelength-positions of the primaries are not ascertained.

9.4 Color models, color spaces and color palettes

A *color model* is the way color is mathematically described with 1, 2 or 3 primaries. A color space is the set of all the colors that can be represented using a specific color model.

---

[122]More precisely: a subset of visible colors that the device can display, or the area enclosed by a color space.

[123]*Beauty is in the eye of the beholder*, quoted by Marcia in *Molly Bawn*, Margaret W. Hungerford (1878).

Once a color system is specified, the amount of each component needed to create a color is given as a number from 0 to 255, and colors become coordinates in that specific color system. Dark corresponds to (0,0,0) and white has three components with value 255. Any other color with three equal coordinates corresponds to a shade of gray (see also Fig. 48).

From the previous Section we know that:

1. a human color space is described in three dimensions;

2. the exact function of the primaries is not critical, as long as they are different and not co-linear, and

3. the color-mixing process can be additive (red, green and blue for video and computer monitors) or subtractive (cyan, magenta and yellow for printing).

The remainder of this Section briefly describes the color spaces that are most important for scientific writing, *viz.*, the generic human-observer color space (XYZ), the color space of computer and video screens (RGB and $sRGB$) and the color space of our printed scientific writings ($CMYK$). At the same time, the concept of *color palette* is introduced: a color palette is a subspace of the color space, and is defined by the graphing software's options.

9.4.1   The CIE XYZ color space

The CIE[124] XYZ color model was developed in the 1930s as a perceptual color space that approximates the human visual response to different colors – that is, two colors with same XYZ will look the same to a CIE standard human observer.[125]   Figure 32 shows the CIE xyz standard observer functions. The CIE XYZ color space contains the human gamut of all perceivable colors, and is mostly visualized as a projection x, y, Y. The Y component (called the luminance of the color) characterizes the intensity of light, and the chromaticity coordinates x, y (defined by the dimensionless ratios $x = \frac{X}{X+Y+Z}$ and $y = \frac{Y}{X+Y+Z}$) stand for an objective measurement of the color of an object, leaving aside all brightness information.

XYZ colors with the same chromaticity values have the same hue, but can differ in absolute brightness Y. Hence, colors can be described either by XYZ coordinates or by xyY (chromaticity plus luminance Y). Figure 33 shows the x, y *chromaticity diagram* (for a given Y). The straight line connecting the two end points of the horseshoe curve is referred to as the *purple line*, which corresponds to the singularity in the Newton wheel when leaping counterclockwise from red to violet. The point of equal energy is the CIE standard for white light. The pure (*i.e.*, fully saturated) colors lie on the ridge of the curve, and any line joining any two points in the diagram defines all colors that can be obtained by combining these two colors.

In daily life there is not much reason to worry about the XYZ system: its role is, just like halftoning, embedded in the graphical rendering software.

---

[124]*Commission Internationale de l'Éclairage*, organization that develops color standards: www.cie.co.at/

[125]The XYZ primaries are not material primaries, but invisible mathematical concepts, because they lie outside the gamut of the real colors.

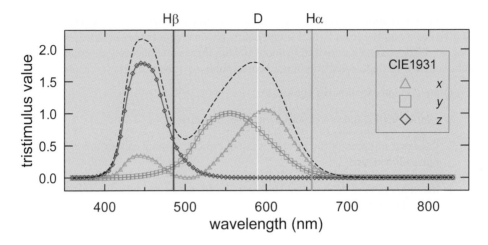

Fig. 32. CIE *xyz* standard observer functions, and their sum (dashed line). The top *x*-axis gives the approximate positions of the spectral lines shown in Figure 31. Both wavelength scales are linear.

9.4.2   The *RGB* and *sRGB* color spaces

The *RGB* color space is a color model for screen colors, *i.e.*, colors related to the emission of light; each number in the range 0–255 represents the intensity in each channel of the additive primaries. The *RGB* gamut is a triangle (Fig. 33) in the *x*, *y* plane.[126] The *sRGB* variant is a standard color space (designed for a typical CRT display) that has become the default color space for WINDOWS. The *RGB* triangle envelopes the gamut of colors that can be obtained by combining the three primaries,[127] but its gamut does not enclose all visible colors. Note the differences with Maxwell's triangle and gamut shown in Figure 30.

9.4.3   The *CMYK* color space

The cyan–magenta–yellow–black *CMYK* color space is not used for emission-light colors, but for reflected-light colors when printing on paper, cloth or other stuff. Cyan, in the subtractive color system, is one of the three primary colors, in the additive system it is a secondary color. Cyan absorbs red, Magenta absorbs green, and Yellow absorbs blue. The reason for adding the black *K* ink is that the subtractive primaries *C*, *M* and *Y* do not entirely remove their complementary *R*, *G* and *B* colors, hence their mixture produces a brownish tint.

The five-sided polygon in Figure 33 shows the gamut of the *US Web-coated SWOP*[128], a generic *CMYK* color profile. There are, however, additional factors to be taken into account, *viz.*:

---

[126]*i.e.*, for a given *Y*
[127]Only the relative amount of each primary is needed to quantify the color.
[128]SWOP: Specifications for Web Offset Publications.

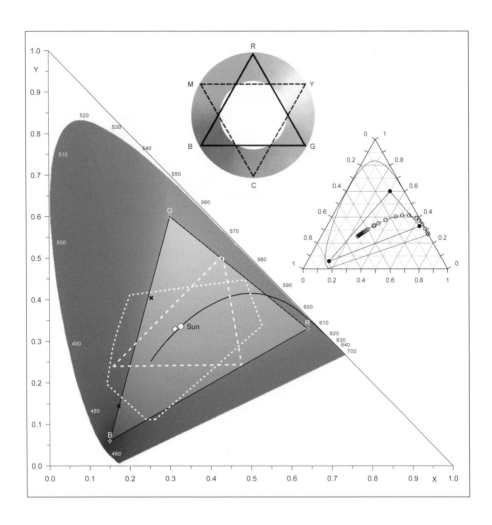

Fig. 33. CIE *xy* chromaticity diagram for the average person, with the *RGB* gamut triangle (typical color CRT monitor) and approximate *CMY* gamut (for color printing). The five-sided polygon is the gamut of a generic *CMYK* color space. The curved black line is the locus of black bodies (500–25 000 K). The pure (fully saturated) colors lie on the ridge of the horse-shoe curve. The position of the Sun is indicated, as well as the white point for a CRT monitor (◇: point of equal energy corresponding to the CIE standard for white light). × represent both A&A blue colors in *RGB* (coordinates from Table 7). The triangular inset shows some of the same data elements in a classic ternary diagram triangle, with some positions of black-body radiators (○) in the stellar-temperature range up to 25 000 K. The colored circle is a rendering of the color wheel representing the visible light spectrum with the positions of the *RGB* and *CMY* colors (Source: Kodak Handbook 1979). Primary colors are separated by 120°, and complementary colors are located directly opposite each other. The colors inside the horseshoe area were adapted from Steer (2004), with permission.

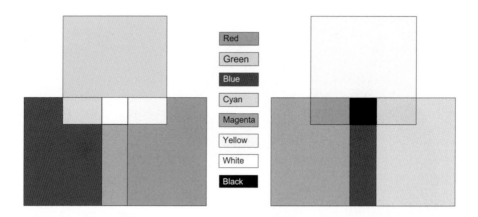

Fig. 34. Primary additive *RGB* (*left*) and subtractive *CMY* (*right*) colors. One graph follows from the other by permutation of colors and swapping white to black.

1. the reflective properties of the print substrate play a significant role: printing on yellowish paper, for example, will yellow the whites, and modify all colors;

2. the environment lighting plays an important role: ambient light of a different color will change the perceived hues – see the horizontal band in Figure 48.

CMYK is very device dependent, and covers fewer colors (a smaller gamut) than *RGB*, with the consequence that **some *RGB* colors cannot be represented in *CMYK*.** Though *RGB* files contain more color data than *CMYK*, they nevertheless measure less in disk space.

The A&A cover and logo uses two hues of blue – these are also used in the timeline shown in Figure 5, and in the central square in Figure 2. Table 7 lists their *CMYK* values and corresponding *RGB* coordinates (with their positions marked in the chromaticity diagram of Fig. 33).

Table 7. The A&A cover and logo uses two hues of blue (no *R* involved).

Color	*C*	*M*	*Y*	*K*	*R*	*G*	*B*	Examples
light blue	72	0	23	0	0	188	201	Figure 5
dark blue	100	65	23	15	0	83	130	Figures 5, 17 and 33

There are basically two types of inks for printing: dye-based inks (*i.e.*, the more common types), and pigment-based inks (which are chemically stable over many decades, are much more *lightfast*[129] than dye inks, and are also much more expensive).

---

[129]Lightfastness is the degree to which a print resists fading when exposed to light.

9.4.4   Device-independent colors: spot colors

When colors cannot be made by mixing primaries, the graphic arts industry uses *spot color*[130], *i.e.*, industrially prepared special inks that are applied one by one. The PANTONE[131] Matching System, for example, uses 14 base pigments to produce each of the more than 1000 spot colors in their list, using a proprietary and unique ink mixing formula.[132] This proprietary color space assures a perfect match of the colors on the customer's monitor with those on the vendor's screen. The lists of color numbers and pigment values are the intellectual property of PANTONE and are thus copyright-protected, therefore they are only provided within high-cost software packages.

9.4.5   Degenerate color spaces

Section 9.3 referred to Maxwell's papers that describe in detail all the intricacies of color vision using black-and-white graphics: he just had no choice, because color printing was simply not available. Society, in those days, was not so colorful as it is now: except for the colors in nature, and the colors in paintings, tapestries, buildings and clothes, people were not much exposed to color. Nowadays, color is ubiquitously used in projection, in monitor viewing and in print, and quite often more for purposes of coding than for aesthetics.

All previous Sections dealt with trichromaticity, the three-dimensional color model in which every color stimulus can be described with three numbers referring to three distinct primary colors. The extreme apex is the *black-and-white color space*, with only two levels of *shade*. Then comes grayscale, a monotone space with 255 gray levels, followed by duotone, a mode with just two colors where the color triangle becomes singular.

On page 126, an allusion was made to persons with red-green color blindness. Such dichromatic[133] form of color blindness is called *deuteranopia*, from the Greek expression for "second does not see", referring to the second cones in the retina (the G channel) that are defective. This does not mean that green colors are invisible to people suffering from deuteranopia: most people with this condition can see quite well in the middle region of the spectrum, though they do not discriminate color differences. In about 5–6% of the (mainly male) population, at least one of the receptor types is affected with this anomaly, which means that about 7 to 8 students that attended the first and second SWYA schools are expected to be affected in one way or another. Such reduced color vision is described by a two-dimensional color model.

Maxwell was not the first to comment on different observers' abilities to discriminate colors. John Dalton (1798), already in 1794 remarked[134] that, unlike persons in general who can distinguish six kinds of color, he can see only two or at most three colors: yellow,

---

[130] An ink that is printed in a single run, like the individual C, M, Y as opposed to *process color*, which combines primary inks.

[131] PANTONE – founded in 1963 – is the developer of the most widespread color communication standard in the printing industry, and is a leading source in the technique of accurate communication of color in printing, publishing and textile industries.

[132] Instead of words or coordinates for describing colors, standardized labels and numbers are used (for example PMS 288, a bluish tint).

[133] A dichromat is a viewer with one of the *RGB* primaries missing.

[134] His lecture was read on October 3[1st], 1794 (note the confusing format of this date).

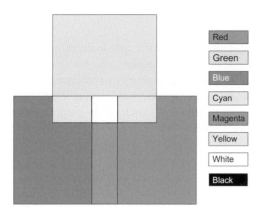

Red

Green

Blue

Cyan

Magenta

Yellow

White

Black

Fig. 35. Simulated *RGB* colors as perceived by a viewer suffering from protanopia. Black and white were untouched, as well as the labels in the legend.

blue and purple, and that yellow and green seem to him one color (for a full citation, see the caption of Fig. 36). Therefore, *Daltonism* is the term referring to dichromacy characterized by a lowered sensitivity to green light, *i.e.*, by deuteranopia. When the red cones are defective, then reds, yellows and greens are indistinguishable, and this deficiency is called *protanopia*. *Tritanopia* – the inability to see the third color (*B*) is quite rare.

Figure 35 reproduces the left panel of Figure 34 with *RGB* colors transformed[135] as if perceived by a protanope. The large majority of readers (about 94% of the population) will thus recognize the importance of selecting appropriate colors when designing graphics. This consideration is not a matter of scientific writing alone, but also a major preoccupation in business graphics: Cole (2004), for example, mentions having received an enquiry from a major bank that sought advice on the use of color in their graphs, because a customer with defective color vision had complained of his inability to distinguish color-coded data in their reports.

Web designers have defined a color palette of 216 colors for use in 8-bit (256-color) graphics mode – called the *Websafe palette* – and other colors are then dithered[136] to that color palette. The scheme is based on six steps of each of the *RGB* display ranges, hence the *RGB* coordinates of each color in this palette differ from the coordinates of each other color by 51 units, or by a multiple of this value. Figure 36 shows the positions of these colors in an *RGB* chromaticity diagram. The first aspect that draws immediate attention is that, though the color coordinates are regularly spaced (*i.e.*, relatively equidistant) in the *RGB* lattice, the spatial equidistance is not conserved in the ternary diagram.

Figure 36 shows that the color perception of people suffering from deuteranopia is very close to a duotone color space. Hence, **your graphical message will stand out when you avoid the confusion locus – common to deuteranopia and protanopia – along the *RYG* line of the spectrum, and when you select blue or white hues** (*e.g.*, for the backgrounds of powerpoint presentations).

Dalton's citation, partly mentioned in the caption of Figure 36, though, comprises a most confusing error in the subsequent sentence: *"The difference between the green part*

---

[135]Transformations based on information provided by the *British Telecom* website, http://www.btplc.com/
[136]Dithering: randomly juxtaposing pixels of two colors to create the illusion of a third color.

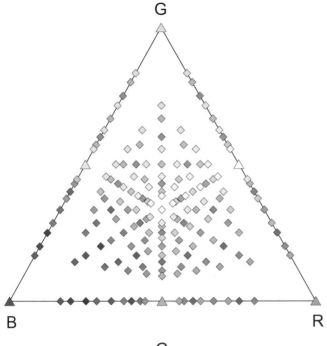

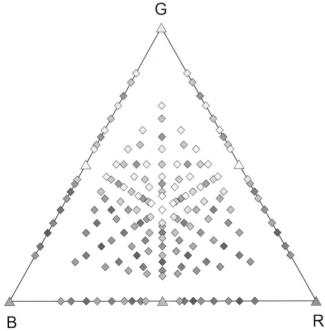

Fig. 36. *Top*: WEBSAFE palette of 256 regularly-spaced colors. The △ symbols for the *RGB* and *CMY* primaries are larger than the ◇ for the composite colors. Black, and the shades of gray (20%, 40%, 60%, 80%), are not shown, since they coincide with the white point with coordinates (1, 1, 1). *Bottom*: simulation of WEBSAFE colors as perceived by a deuteranope, accompanied by Dalton's *verbatim* description of his own color gamut.

THAT PART OF THE IMAGE WHICH OTHERS CALL RED, APPEARS TO ME LITTLE MORE THAN A SHADE OR DEFECT OF LIGHT; AFTER THAT THE ORANGE, YELLOW AND GREEN SEEM *one* COLOUR, WHICH DESCENDS PRETTY UNIFORMLY FROM AN INTENSE TO A RARE YELLOW, MAKING WHAT I SHOULD CALL DIFFERENT SHADES OF YELLOW (DALTON 1798).

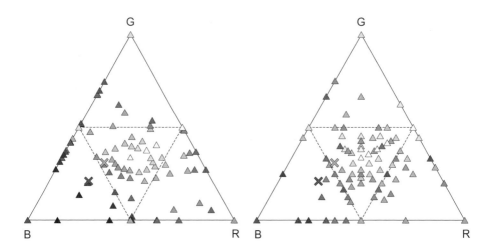

Fig. 37. Two standard color palettes. *Left*: PHOTOSHOP, *right*: GRAPHER. The × signs indicate the positions of the "A&A blues". Axis tick marks and labels have been omitted. Each apex of the triangle has the corresponding tristimulus value 1 with the two other values 0.

and the yellow part is very striking to my eye: they seem to be strongly contrasted." – he of course meant to say "between the green part and the blue part", a regrettable *lapsus*[137] most certainly overlooked during shallow proofreading.

9.5   Color workflow

9.5.1   Color management

Most graphical design software will open in the s*RGB* default color space, which is quite poor for evaluating the printed outcome. As an alternative, an *RGB* working space with a maximum overlap with *CMYK* can be selected. The monitor must be tuned to a contrast setting that is as close as possible to print, by applying an appropriate *gamma correction* that characterizes the nonlinear relationship between image pixels and monitor brightness. Gamma (Γ) is a parameter that assigns the relative brightness (input to output) of midtones: when $\Gamma = 1$, there is no adjustment (*i.e.*, the default), with $\Gamma > 1$ the dark areas receive more detail, with $\Gamma < 1$ the light areas get more tone resolution. Γ variation is nonlinear, as it conserves the responses at the white and black ends. For optimal prepress work, a gamma of 1.8 should be selected. In other words: **a color space always includes a gamma specification**.

Color management is the control of all aspects of color from the color space of the conception and design stage, up to the color space of the final printing – again, I must refer

---

[137]A slip or lapse.

here to Shannon's model in Figure 2. Color management systems try to take care of the different color reproduction characteristics of monitors, cameras, printers and scanners, so that the colors look the same, regardless of the hardware or the operating system used to reproduce them. Without color management one cannot hope to accurately predict output color on the basis of input color. Color management relies on ICC[138] profiles (file extensions .icm or .icc), which are files that contain the data of the color characteristics of the output device. This information describes how the scanner, monitor or printer reproduces input colors.

An element of the color management is the choice of the *color palette*, a subset of preselected colors provided by your graphical software: finding the right color palette is almost as important as drawing the graph. Figure 37 illustrates two such standard palettes in a ternary *RGB* diagram. As the graph illustrates, both examples cover different areas and color ranges, and even reveal color anomalies. One can usually extend the palette through the use of a *color picker* that allows you to select any *RGB*-additive mix.

9.5.2   Proper color balance and gray tonality

The Publisher/Printer can adjust color balance and exposure only if you provide calibration exposures. The front cover illustration, for example, features a *Kodak Color Separation Guide and Gray Scale*, a reflection step chart of 20 stepped, gray values[139] for comparing tone values of the original with the reproduced image (see Fig. 38, and also the back cover). The color-control patches strip contains standard colors used in *RGB* and *CMYK* to help match the colors in the printout with those of the original. When such an exposure is included in a photographic reproduction, it is possible to faithfully render the original colors in the printed image.

Let us now see how the technical principles apply to real life. Figure 38 shows the calibration strips for the cover illustration in *RGB*. The Publisher's prepress adjusts the grayscales and transforms the *RGB* to *CMYK*, applying in-house color and grayscale profiles that support their calibrated printers.[140] Figure 39 shows measured brightness level (on the left y axis) and normalized pixel value (on the duplicate y axis) as a function of reflectance or density[141] for three different exposure types: the middle grayscale shown in Figure 38, a sample scanned Kodak grayscale, and the Kodak grayscale photographed together with the book cover illustration. The two lines have different slopes, which means that applying the upper curve will result in an image with less contrast and lower dynamic range,[142] than when applying the lower line. The shape of the Kodak grayscale curve is nonlinear as a result of using a nonlinear detector, the photographic film.[143]

---

[138] *International Color Consortium*, http://www.color.org

[139] 20 steps in 0.10 density increments ($\frac{1}{3}f$-stop) between 0.0 [white] and printing black. Label 0 (A) corresponds to density 0.05, label 19 corresponds to 1.95.

[140] Note that the cover and the pages of the book are not printed on the same printers.

[141] Density equals $-\log_{10}\frac{\text{reflected light}}{\text{incident light}}$, and density 0.05 is the reflectivity of white paper ($\sim 90\%$) because typical paper increases density by about 0.05.

[142] The dynamic range of an image is the range between the lightest and darkest pixels in the image. Beware that recordable dynamic range and displayable dynamic range can differ.

[143] This curve also embeds a basically linear component due to the scan of the photographic positive print.

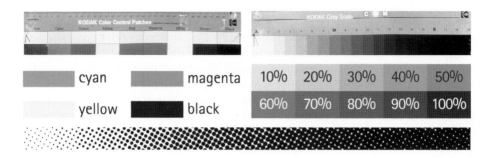

Fig. 38. *Top left:* Kodak control color patches for the cover illustration. *Top right:* Kodak grayscale strip. *Middle:* *CMYK* patches and gray scale used for printing the cover (both middle images are also reproduced in full color on the back cover of this book). Note that the labels for densities 10–50% are dark on a relatively bright background, and *vice versa* for the labels 50–100%. *Bottom:* halftone image of the Kodak grayscale strip, maximum dot size 7 pixels, screen angle 120°.

The point I make here is not so much the technical aspect by itself, but the fact that the author or designer (leftmost box in Fig. 2) has not the slightest control of what the reader (rightmost box in Fig. 2) will see if there is no **accurate communication with the Publisher's prepress staff**: this connection thus becomes one of the greater sources of *communication noise* alluded to on page 69. Just check for yourself: compare Figure 27 (in the color pages) with the reproduction on the back cover of the book – at the moment of writing, I had no control whatsoever on the degree of color matching of both prints.

9.5.3   Color transformations

But if every person and every output device has its own color space, how can we ever hope to communicate, especially since all these spaces evolve with time?[144] Just like language translation, color conversion is not always a straightforward one-to-one mapping.

The first type of transformation is related to the person who views the screen. Such transformations consist of matrix multiplication plus vector translation of the form[145]

$$
\begin{pmatrix} R \\ G \\ B \end{pmatrix} = \begin{pmatrix} 3.2209 & -1.5392 & -0.4979 \\ -0.9696 & 1.8762 & 0.0408 \\ 0.0554 & -0.2047 & 1.0573 \end{pmatrix} \begin{pmatrix} X \\ Y \\ Z \end{pmatrix}
$$

Each monitor has its own linear *XYZ*-to-*RGB* transformation matrix, and bringing two adjacent *RGB* monitors to the same standard involves multiplying the color matrix of one monitor with the inverse color matrix of the other. When color-calibrating monitors and printers, you should keep in mind that

---

[144]The eye ages with time, computer screens age with time, even ambient temperature plays a role.
[145]The coefficients of this matrix are just one possible set of coefficients.

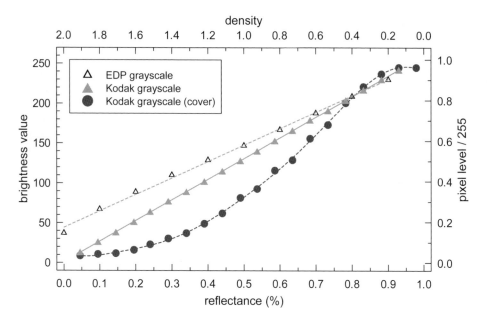

Fig. 39. Brightness as a function of reflectance or density for three different exposure types. EDP grayscale △: measured pixel level as a function of reflectance for the middle grayscale in Figure 38 (reflectance data provided by the Publisher of A&A). Kodak grayscale ▲: pixel measurements from a scanned Kodak grayscale as a function of density. •: pixel values from the Kodak grayscale photographed together with the book cover illustration as a function of density (see text). Curves are linear and polynomial fits. Note the symbol sizes in this graph: triangles and circles all have the same nominal 8-pt size, yet they are not perceived as of equal size.

1. the above relations lead, mathematically speaking, to a perfect fit as soon as three independent tristimulus values are provided. To be robust, however, many more empirical data must be acquired in order to secure accurate color coefficients from a least-squares solution;

2. the entries should have a reasonable extent of color diversity: calibrating with a handful of points that are close in the RGB space will always yield a poor calibration with a very limited range of validity (just like in astronomical photometry as referred to on page 104);

3. monitors age with time, and their gamuts change along, hence a good calibration setup of today will not necessarily work out well one month later, and

4. when adjusting several monitors to the same scale, each monitor must be separately calibrated. A matrix with coefficients that are averages from several other matrices, will never be a proper calibration.

As mentioned already, RGB has a larger gamut than $CMYK$. Therefore, although one can convert any RGB to $CMYK$ without losing information, the inverse is not the case:

when a *CMYK* color is converted back to *RGB*, information is lost. For the conversion to *CMYK*, the A&A Publisher uses the *ISO coated V2* profile[146] to generate the PDF file of the illustration that is printed on the cover of each Volume of A&A (Hosotte 2009).

It is very important to understand what happens to your graphics during the journey from your desk to a digital or printed publication. The process is not very widely understood, even by people who work in the printing and in the graphics industry. For Sterken (2009), I asked to print the same 12th-century world map on the cover, and also inside the same book. At proof stage, I expressed my unhappiness about substantial color differences between the prints of the same map. The technician told me that it was *"impossible to equalize the colors since the production involves two different printers"*. The problem really was that one person had transformed my manuscript to a *CMYK*-based PDF file, whereas I had delivered the cover as an *RGB*-based PDF. A publishing or printing company with a good prepress service should anticipate such situations, moreover the color profiles of their respective printers should be so well maintained that all their equipment operates in concert, without the client even having to think about the issue. The really good companies calibrate their monitors and printers on a weekly basis. Unfortunately, *canned* profiles – generic profiles for a specific brand and model – are too often used instead of custom profiles.[147] This example illustrates, again, the role of the Publisher's prepress services, as described in Section 3.2 in the paper by Agnès Henri in Part 1 of these Proceedings.

9.5.4   When to change color space?

The basic workflow, in principle, is to design all artwork **in the color space with the largest possible gamut**, and to convert to a more restricted gamut at the print-ready moment only, *i.e.*, **make all retouches and changes always in the same wide-gamut system, and transform only when the job is ready to be sent to the printer**. Moreover – since *RGB* files cover a larger gamut than *CMYK* files, at the same time they weigh less in megabytes – keeping an *RGB*-based workflow is the most versatile and the most economic approach. Using the language metaphor on gamuts (page 130), it is obvious that making repeated changes by **transforming back and forth between color systems is like translating the same sentence back and forth between languages**: every translation destroys information, and increases communication noise.

The commercial printer at the end stage uses profiles to convert your *RGB* space to a very specific *CMYK* color space that takes into account the press, the inks and the paper stock. This transformation should happen only once: **at the very final stage, where most of the money is spent**, and this transformation is done by the company that prints your work, in the same automatic and invisible way as the halftoning is performed. The only way to evaluate the color print, is by proofprinting on the final device, which is a costly affair. Alternatively, a well-calibrated printer will do, and a lot of trouble can be avoided by "proofing" via previewing on screen in *CMYK* mode.

---

[146]ICC standard color profile created by the *European Colour Initiative* (ECI) www.eci.org

[147]Not to speak of variations in color that are caused by variations in ink supply, room temperature and humidity, the batch of paper, and whether the print run has just commenced, or is at its end.

10   Principles of graphical excellence

Edward Tufte published a series of exquisite books (Tufte 1983, 1990, 1997) on graphical design, with some of his material presented already for the first time by Funkhouser (1937). It is most interesting to compare the print quality of Tufte's and Funkhouser's publications, and see the enormous advances in print technology over half a century, in all aspects of resolution, font and (lack of) color. Though the content of these books and papers is only partially useful for astronomy communication, the general principles are valid for about all aspects of our graphical work. Tufte (1983) phrases the principles of graphical excellence in the following terms:

> GIVE TO THE VIEWER:
> THE *greatest* NUMBER OF IDEAS
> IN THE *shortest* TIME
> WITH THE *least* INK
> IN THE *smallest* SPACE

In other words: optimize in all possible directions in order to obtain the highest possible information content characterized by clarity, precision and efficiency. He also formulates five rules:

1. above all else show the data;

2. maximize the data-ink ratio;

3. erase non-data ink;

4. erase redundant data-ink, and

5. revise and edit.

By way of experiment, I checked all graphics of the first three years of A&A (1969–1971) and compared them with graphics appearing during the period 2007–2009. Although the recent graphs look a lot more professional than those of the early days, some classes of errors commonly made in the 1970s are still committed today! In the subsequent Sections I concentrate on some of these errors.

10.1   Managing ink

Tufte defines the *data-ink ratio* as the ratio of data-ink to the total ink present in the graph, *i.e., the percentage of ink used for displaying non-redundant information.* In particular, thick grids, crosshatching, dark shading, labels and ticks, oversize legends and rubble of computer plotting (non-specified characters or identifications[148] at the top or bottom of a graph) decrease this ratio, and render graphs unaesthetically overladen. In other words: **whatever ink you add to a graph should have a reason**. This also refers to his concept of *data density*.

---

[148]The kind of identifier, also known as chart junk.

$$\text{DATA-INK RATIO} = \frac{\text{DATA-INK}}{\text{TOTAL INK PRESENT IN THE GRAPH}}$$

$$\text{DATA DENSITY} = \frac{\text{NUMBER OF ENTRIES IN THE DATA BASE}}{\text{AREA OF DATA GRAPHIC}}$$

Figure 40 is a typical example of a vintage graph hand-plotted on a printed polar-coordinate grid. Here you see that the heavy grid yields a very low data-ink ratio because of the closely-spaced grid cells, and because of the substantial weight of the grid lines. Moreover, the data area encompassed by the three line graphs identified in the figure legend, is less than 15% of the complete graph area. The "above all else show the data" principle could be implemented by restricting the graphics to a rectangular part delineated by the superimposed central square that I added for the sake of argument. The purpose of the square is also to illustrate an optical illusion that comes along when applying heavy raster lines: the sides of the square – although strictly rectilineal – appear slightly curved when viewed on the background of the circular grid.

Another example is shown in Figure 41, based on data kindly provided by Chanut *et al.* (2008), illustrating the temporal evolution of the longitude of a planetesimal. The left diagram shows the complete dataset. The huge block of ink covers more than 100 000 datapoints, yet to the reader's eye it is only one single pixel of information that covers 65% of the graph. Since the dependent variable is an angle, a non-data line is automatically added each time the argument passes from 360° to 0°, and a substantial amount of non-data ink is thus added. Nevertheless, it is not easy to find a good solution to present such a cyclic dataset: a polar graph offers some relief, but is not a really good method to help the viewer understand the process. One possible remedy is depicted in the right panel of Figure 41: applying an axis break removes the bulk of the disturbing ink, at the same time this gives more focus to the real behavior of the independent variable "trapped" in a confined interval after $t = 65\,000$. Remember, though, that a poorly-documented axis break may lead to confusion: in this case the eye may not readily recognize the removal of an interval of more than 60 000 years, although in Figure 41 the scale problem is documented by including proper tick marks on both sides of the interrupted axis. Graphs with similarly high ink density are very common when displaying frequency spectra of cyclic or periodic data.

The effect of non-data ink depends on the situation: Figure 18, for example, is quite clear when viewed in color mode, but its grayscale rendering of the non-data ink is somewhat confusing.

10.2   Revise, edit, revise, edit, and revise, edit

Revision of graphics is a most important act, that very well matches the phase of revising and proofreading texts as discussed in Section 10.1 in Paper I. Let it be clear: the first draft is only a working model – just like in the writing process – and the subsequent

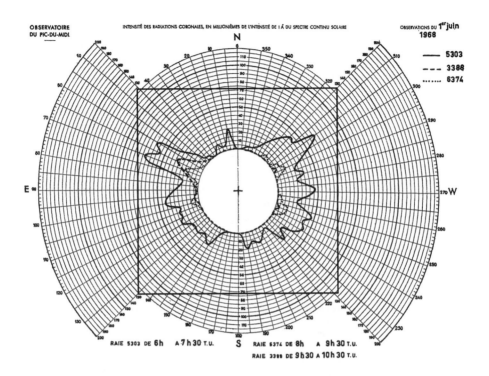

Fig. 40. Vintage graph hand-plotted on a printed polar-coordinate grid. The central square has been added for emphasis (see text). The title and legends provide, besides identification of the data, also information on the time and location where the data were obtained. Based on Rozelot (1969).

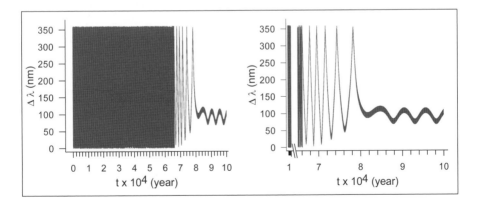

Fig. 41. Examples of graphics that is difficult to read. *Left*: the complete dataset. *Right*: same, with axis break. Duplicate axes were omitted for clarity. Line color for this graph was set to blue, in order to obtain a gray rendering in black-and-white print. Based on data from Chanut *et al.* (2008).

drafts are for optimizing and for colorizing. There is an absolute need for multiple drafts: experiment with several types of graphs, play with colors, change aspect ratios, and so on. After the last revision, a thorough "proofreading" of all graphs is necessary, with special attention to all elements of Section 7. A special act of proofreading is to verify how the color graphics appear in grayscale by converting the graphs to monochrome (or by setting the dvi viewer to black-and-white, or simply by printing all graphs on a black-and-white printer). Paper I referred to **proofreading for content** and to **proofreading for errors** – likewise proofreading graphics comprises both aspects, and your last check before submitting a manuscript for publication, should equally be a deep proofreading of all graphics for errors and for inconsistencies.

Extreme care must be observed with respect to the validation of any tabular material that was imported through scanning and subsequent OCR, especially when data are extracted from documents that date from before typesetting by computer was implemented. Such OCR is not without obstacles, as various defects may deteriorate the data. The following problems regularly occur:

1. when dealing with century-old books and catalogs, the user is not always allowed to xerox or to scan valuable documents, and hence must work with image files from online libraries or repositories, and such scans are often of too low resolution to adequately apply OCR;

2. older manuscripts (for example, Maunder 1904) had to save on typesetting labor, hence leading decimals and + or − signs were given at the beginning of a column only, unless they change midst a column, and only then the change was indicated;

3. confusion between o, 0 and °, between l, [, ! and 1, between 5 and 6, etc.;

4. obsolete characters – such as the ø for 0 – leading to grave errors;

5. extraneous spurious characters resulting from blobs of ink on the paper, to specks of dirt on the scanner's glass plate;

6. broken letters and symbols, for example part of a 0 may be missing and the character is then read as (, or a missing segment of 6 may lead to recognition as 0;

7. columns with missing or non-available data are sometimes filled with blanks, or with a sequence of dots, rendering such OCR products not strictly machine-readable;

8. genuine typographical errors by the typesetter, and missing characters (especially the decimal dot).

10.3   Layout: the typeset page as an image

Tufte (1983) says *"data graphics are paragraphs about data"*, hence the placement of these paragraphs is of prime importance: a typeset page, when considered in its entirety, simply is a composite image compiled in a semi-automatic way. Because we look at a printed page in one glimpse, we also see the aesthetics of the graphics paragraphs and the text paragraphs at the same glance. But each page can display several unaesthetic and even disturbing artifacts, as described in the list below. In addition, Section 13 in Paper III also discusses the possible effect of page layout on the *interpretation* of the text (*i.e.*, the case of Robert Millikan's alleged data cooking).

1. **Widows and orphans** are dangling words at the beginning or end of a paragraph. A widow is *a short line appearing at the top of a page, and is at the end of a paragraph*; an orphan is *a short line at the bottom of a page, or a word or part of a word on a line at the end of a paragraph.* Figure and Table captions can also produce such unaesthetic dangling words. Such artifacts result in excessive white space between paragraphs or at the top or bottom of a page, and can be avoided by changes in wording. I am often astonished to see widow-type appearance of one or two references appearing on a blank page at the end of a paper. Such occurrences involuntarily create the impression of an extra page (especially in proceedings books where space is limited), though they can so easily be avoided by technical tricks (slight resizing of one Figure, using abbreviations for journals, etc.).

2. **Alignment and right-margin justification**: *right justified* is often used instead of the term *right aligned*. Right justified is type aligned both right and left, right aligned is flush right and ragged on the left. Two-column page layouts often lead to unaesthetic alignments with lots of white space; an example is shown in the caption of Figure 13.

3. **The aspect ratio of graphics** (as already discussed on page 118) has two facets: the internal aspect ratio, *i.e.*, the optimal ratio that shows the data in the best way, and the page aspect ratio, as defined by the Publisher's rules and guidelines. It is the latter that causes most of the unaesthetic page layouts in journals. Since Publisher's rules cannot be changed, the internal aspect ratio should be optimized.

4. **The composition of graphical elements**, as illustrated in Figure 36. The most straightforward approach would be to place the triangles next to each other, but that requires downsizing, and results in loss of color perception. The more aesthetic solution is vertical arrangement above a caption over the full width of the page. The actual presentation is perhaps less pleasing at first sight, though it permits a direct explanation, in John Dalton's own words as published more than two centuries ago.

10.4 Style

Paper I stated that, in contrast to prosaic style, *"scientific style is direct, plain, assertive, context-independent, and brief ... "*. The same applies to graphics: your graphs will develop in a certain style direction, just like architects' houses bear the signature of their designers. That is why it is important to develop a good style as soon as possible, and I refer to Section 13 for a list of errors to avoid. Your graphical design skills will equally benefit from looking at all aspects of graphics that you encounter in books, journals – even in the daily environment around you. The most challenging task is to **remain consistent inside the same paper**, a goal that is quite difficult to achieve in a multi-authored work.

11 Lying with graphs

Some images, by definition, always "lie": *false-color*[149] representations of celestial photographs do mislead, especially images taken in non-optical windows of the electromagnetic spectrum. For example, only part of the Hubble Space Telescope wavelength window overlaps with the gamut of human vision, and other space telescopes even observe
only in the ultraviolet or in the infrared spectral region. All the same, infrared space images acquired in a 75-μm band are combined with 150-μm and 300-μm band data and
rendered with selected *RGB* primaries to yield pleasing and informative pictures.

The last paragraph of the Preamble in Paper I stated that we proceed *"on the premiss
that the reader is strongly committed to truth in science, and even more so to the truthful
communication of scientific findings"*. Still, good intentions do not always suffice: graphics may – inadvertently or on purpose – mislead the reader.[150] Images are just ideal for
seducing or deceiving, and that particular power is the core force of all advertisement and
publicity messages. However, an awesome drawing or a fantastic picture can never be the
substitute for a proof or for cogent evidence, as Ronald A. Fisher (1925) so clearly stated:

> DIAGRAMS PROVE NOTHING, BUT BRING OUTSTANDING FEATURES READILY TO THE EYE;
> THEY ARE THEREFORE NO SUBSTITUTE FOR SUCH CRITICAL TESTS AS MAY BE APPLIED
> TO THE DATA, BUT ARE VALUABLE IN SUGGESTING SUCH TESTS, AND IN EXPLAINING THE
> CONCLUSIONS FOUNDED UPON THEM.

11.1 The copyright

The same rules on copyright as explained in Papers I and III apply to the use of graphics:
you must always mention the source, and whenever it is obligatory, ask for permission to
reproduce. This chapter contains several reproductions of copyrighted material with the
copyholder's permission (granted for free, or for a fee). Figures 29 and 30 are copyright-
expired images, since their authors have passed away more than 50 years ago, hence these
images are in the public domain. But remember that, though the tapestry reproduced on
the front cover of this book is more than 500 years old, it is the artwork's present owner
(*Röhsska Museet of Fashion, Design and Decorative Arts*) that must give or sell the right
to copy. Even if this artwork had been published in an artbook, it would still be protected
through the copyright clauses in that book.

A special warning is in place about copyrights of internet images. Many images on
the web have been uploaded without the slightest worry about copyright violations. Nevertheless, you do not have the right to copy and publish any web images without seeking
permission from the copyright owner. Beware that some copyrighted material embeds
well-hidden digital marks that allow tracing back the origins. Some publicly posted files
even have digitally changed content (an intentional graphical bug) that will reveal the
source of the image. Some graphics presented by Wikipedia are licensed under the *GNU*

---

[149]False-color differs from pseudo-color: the former refers to an *RGB* rendering of images acquired with
non-*RGB* filters, whereas the latter refers to a monochromatic image represented in color, see Figure 4.

[150]Wilful deceit with graphs is discussed in Section 6.7 in Paper III.

Free Documentation License,[151] and may thus be reproduced without explicitly asking for permission. Again, **permission to reproduce is not the same as transfer of copyright**.

11.2   Retracing graphs

But what about so-called *derivatives*: are you, for example, allowed to scan Figure 13, and retrace[152] it? And if the retraced version is modified (for example by adding a branch to the tree in Fig. 13), is that enough to claim new copyrights? If you reproduce either a scanned drawing or a traced version, you cannot claim copyright on the image since it is a slavish reproduction. Stronger even, nobody else may reproduce your version, whether or not you forgot to reveal its source. However, if you introduce original variations in content, the situation becomes more complex, take for example Figures 1 and 2, where the former is a slavish copy, whereas the latter is an expanded version with added interpretation. Note that adding only color is never enough for claiming copyright, so Figure 30 in Paper III – though not a slavish copy – is a derivative and thus, not new.

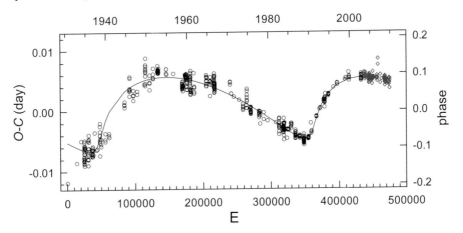

Fig. 42. *O* – *C* curve of CY Aquarii with respect to the ephemeris of Fu & Sterken (2003). The continuous curve is the light-time orbit solution; o: data from Fu & Sterken (2003); ◇: new data. Based on Figure 4 of Tuvikene *et al.* (2010).

One problem with retracing and reproducing is that you should avoid to change the content of the original in such a way that the original message gets distorted. Figure 42 is an example of a reproduction of an *O* – *C* diagram of the pulsating variable CY Aquarii adapted from a graph from Tuvikene *et al.* (2010), which, in turn, is a Figure from Fu & Sterken (2003), overplotted with new data (the ◇ symbols) and an axis extended to fit the original. Note that the four axes of this graph have different units, and that the opposite

---

[151]The *GNU General Public License* is a free, copyleft license for software and other kinds of works: www.gnu.org/licenses/gpl.html

[152]Retracing is a kind of reverse-engineering, in fact the twin of text OCR. In the past, tracing was done on transparent (and often gridded) *tracing paper*.

axes are redundant, but help the reader to interpret the data. The lower x axis is the cycle number (commonly indicated by E), the top axis shows the equivalent time baseline, and reveals that more than 80 years of data are involved (the star's pulsation period is $0^d.0610$, hence the large E numbers). The left ordinate axis gives the residuals in days, whereas the opposite ordinate axis uses dimensionless phase (residuals divided by the period). This axis also gives additional information: it tells the reader that the meandering of the period covers almost 40% of the cycle length. So, Figure 42 barely differs from its source in Tuvikene *et al.* (2010), and cannot possibly be called a new graph, simply because the *added* information content is *nihil*[153]. The Tuvikene *et al.* (2010) Figure, however, significantly differs from the source in Fu & Sterken (2003), and can be considered new – although more than 75% of its content was retraced – because the *added* information content is significant.

11.3   The lie factor

When visual representation is inconsistent with the numerical content, there is distortion. Tufte defines the concept *lie factor* as the ratio of the size of the effect shown in the graph to the size of the effect in the data. Lie factors of 2–5, unfortunately, seem to be quite common.

$$\text{LIE FACTOR} = \frac{\text{SIZE OF EFFECT SHOWN IN THE GRAPHIC}}{\text{SIZE OF EFFECT IN THE DATA}}$$

In particular, one must **make sure that interpretation of graphics does not depend on any variation in graphic design**: only those graphical methods should be applied that do not modify data content, as is stated by Butler (1979):

> . . . IT IS INCUMBENT UPON THE USER TO ASCERTAIN WHETHER THE MANIPULATIONS REQUIRED TO PREPARE THE GRAPHICAL DEVICE HAVE IN ANY WAY MODIFIED THE RELA-TIONS AMONG THE SELECTED VARIABLES.

Auto clipping, auto scaling, auto binning and auto axis-breaking are some of the automatic features that can lead from misrepresentation to outright lies. Remember, though, that some of these variations are often enforced in an automatic way by the addition of a dimension, see Section 4.3.4 on representing 1-dimensional data by 2 or 3-dimensional quantities. Especially fitting a third dimension to histograms and bar charts is like outfitting a 2-D movie with an extra 3-D dimension: it adds nothing to the information content, although it can distract.

---

[153]From Latin, used for "nothing" or "null".

11.4   Outliers

An outlier in a set of data is defined by Barnett & Lewis (1984) as *"an observation which **appears to** be inconsistent with the remainder of that set of data"*, or *"one that **appears to** deviate markedly from other members of the sample in which it occurs"*.[154] These authors underline that the phrasings **"appears to ..."** are crucial: they emphasize that it is a matter of subjective judgment on the part of the observer whether or not to pick out some discordant data for scrutiny. Daniel Bernoulli, in 1777 already, dealt with the problem of outliers in astronomy, and wrote[155]

> IF THERE IS NO SUCH REASON FOR DISSATISFACTION I THINK EACH AND EVERY OBSER-
> VATION SHOULD BE ADMITTED WHATEVER ITS QUALITY, AS LONG AS THE OBSERVER IS
> CONSCIOUS THAT HE HAS TAKEN EVERY CARE.

A similar statement was made by Delambre & Méchain (1806), see Section 6.4 in Paper III.

Some experimental results can just be plainly impossible, but others can appear quite unlikely, though still possible – remember that in a normal frequency distribution 0.2% of the scores may deviate from the mean by more than 3 standard deviations – which makes rejection of outlying values difficult to justify.

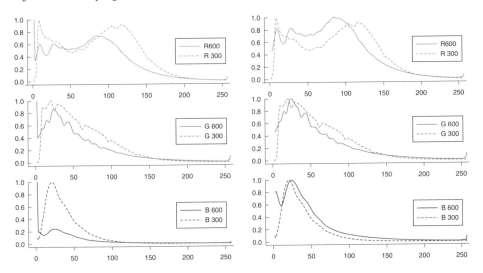

Fig. 43. *RGB* histograms of the cover image based on 300 and 600 dpi scans. *Left*: unedited histogram data. *Right*: after removal of an outlier at pixel value 0 in all channels.

Outliers can easily fool you during the data analysis. Take the odd histogram of Figure 28, and compare it to the same information presented in Figure 43. The left stack

---

[154]The boldface was not part of the original statement, but the emphasis was expressed in the text.
[155]Translation from Allen (1961).

of curves in Figure 43 – actually histograms drawn as curves – are from the same source as the data shown in Figure 28 (the *B* panel actually is identical to the one in Fig. 28). The right stack is derived from exactly the same data, however one outlier – a huge spike at pixel 0 in each band – was removed. The outcome is astounding: whereas *B* shows the largest variance in the left set of graphs, the removal of one point makes both histograms almost identical in those parts of the image devoid of the darkest pixels.[156] The *R* channel, on the other hand, now becomes the most divergent. The single point that was removed, is commonly referred to as outlier or wildshot observation. Barnett & Lewis (1984) distinguish three sources of outliers that can be encountered:

1. inherent variability: the amount of variability belonging to the population;

2. measurement error: inadequacies in measurement superimpose additional variability on the inherent factor (these authors also consider rounding errors as part of the error of measurement);

3. execution error: due to the imperfect collection of the data (*e.g.* the sampling bias and other effects).

But what to do with outliers in your data sample? Let me give three examples of how to properly deal with outliers.

- Appenzeller (1972), when monitoring the luminous blue variable R 81 in the LMC, obtained some magnitudes of the kind that likely would be removed by any automatic outlier-clipper. Tubbesing *et al.* (2002) also collected such "deviant" data, and identified these magnitude drops with eclipses. By combining Appenzeller's published data with their new observations obtained thirty years later, these authors could improve the binary period by a factor of ten, resulting in $P = 74.566 \pm 0.014$ day. Although variability of this amplitude was unexpected, these were outliers that perfectly belong to the inherent variability of the subject. Figure 6 illustrates the statistical properties of these data in a box-whisker plot for three samples where inherent variability is present, clipped, or removed.

- In a CCD-based light curve of a variable star, we noticed a very strange outlier present in one of the photometric bands at only one single moment. After verifying and rejecting all instrumental explanations, a small programming bug[157] was found in the routine for estimating the instrumental magnitudes of the stars: it so happened that only in one single case did the effect become visible in the data. The effect disappeared after the bug was corrected (Tuvikene 2011).

- In Figure 42, two outliers are manifestly present in the data collected since the year 2000: the two data points not only "deviate markedly" (terminology used by Barnett & Lewis 1984) from the data almost simultaneously obtained by other observers, they also deviate from the model orbital solution curve. Similarly large deviations – positive and negative – also showed up in the past, but since these two data points

---

[156]Note that for this exercise the tapestry image was cropped to include only the textile picture frame.
[157]Misidentification of a sharp-peaked point spread function as a cosmic-ray pixel.

were taken from the literature, there is no ground for eliminating them. Two other outliers (not shown) were, however, removed because we found out that the authors had not applied heliocentric corrections to the times of observation taken from an online catalog (see Tuvikene *et al.* 2010, and references therein).

Graphical inspection of the data can reveal outliers, but a box plot of the full range of variation of the data is most useful too: outliers show up as surprisingly high maxima or surprisingly low minima. The discussion of these box-whisker plots in Section 4.2 mentions the relevant statistical quantities, *viz.* the lower (or first) quartile $Q1$ that cuts off the lowest 25% of the data (also called the 25th percentile), the second quartile $Q2$ or the median, which cuts the dataset in half (50th percentile), and the upper quartile $Q3$ that cuts off the lowest 75% (75th percentile). The distance from the first to the third quartile is called the interquartile range: $IQR = Q_1 - Q_3$. Hence, the box plot displays the full range of variation, the likely range of variation (the IQR), and a typical value (the median). See also Sterken (2011) for an application of box-whisker plots to large datasets with outliers.

John Tukey gave a rule for handling outliers: they are either $3 \times IQR$ or more above Q_3, or $3 \times IQR$ or more below the first quartile. For suspected outliers, a coefficient 1.5 is applied: observation m is a suspected outlier if $m < Q_1 - 1.5 \times IQR$ or $m > Q_3 + 1.5 \times IQR$ (see also the Table in the caption of Fig. 45). Outliers should be carefully dealt with:

1. by visual inspection of the original (raw) data – for example, CCD frames should be inspected for mishaps;
2. the distribution of the *control data* – *i.e.*, the comparison, calibration and standardization data – should be analyzed with an eye towards the presence of outliers;
3. the software pipeline should be thoroughly verified.

Finally, take very seriously the advice of Landolt (2007):

> BE SUSPICIOUS OF, BE KNOWLEDGEABLE ABOUT, ALGORITHMS WHICH CLIP ERRANT DATA POINTS. LOOK AT ALL DATA INTENSELY. YOU MAY BE THROWING AWAY YOUR NOBEL PRIZE!

By all means, you should never automatically delete an outlying point, and if you decide not to show such a point in a graph, the Figure caption should state that there was one point out of scale. It is sad, but outliers are just too often shoved beneath the rug.

11.5   Dot joining, or "helping" the eye by connecting points

Plotted data points are often joined by lines, either because it is the graphing software's default option, or because the designer decided that the points must be linked, either for visual clarity (like in Fig. 13), or because there is a functional basis for doing so. Dot joining is often applied to mask *missing data*, i.e., when no data were collected (for example, because of daytime interruption in the observing window), or when the data are simply not good.

Note that when datapoints are connected, the graph may give the impression that interpolation in between points is allowed, which is not always the case if the data gathering was not a continuous process. Figure 19 is such an example: the first line segment for spectral type A0 may not be used for interpolating color indices, because there is no information whatsoever about what the impact of the Balmer discontinuity on the color index is. In the same vein, Figure 1 of Paper III shows connected journal impact-factor data, although this bibliometric index is released on a year by year basis only. Figure 14 is an example where a bar chart can also be represented as a line plot: since there are no data in between the bar values, interpolation is not justified, nevertheless the left panel version of the data turned out to be a more convincing presentation than the right panel, a confirmation that dot-joining fills data gaps and fools the eye.

Beware also that the eye has a natural tendency to see patterns that in reality do not exist. A telling example is the famous story of the *canali* observed by Giovanni Schiaparelli during the Mars opposition of 1877, and confirmed by later observers – though not by every one. Evans & Maunder (1903) consequently set up experiments to find out whether the impression of a network of lines could be produced during observations of disks of diameters 8 to 16 cm, seen from distances of 5–11 m by a class of schoolkids. These authors concluded that the eye sums up the details that it cannot resolve, and that canal-like impressions seem to arise by the *"tendency to join together minute dot-like markings"*. Similarly, the mind sees data trends that are not present,[158] and hand-drawn curves that bridge the gaps "to help the eye" – typified by spurious inflection points and excessive convexity – are very convincing ways to tell the reader what is in the mind of the author, and not what the data say (as can be seen in Fig. 10). This is especially true when estimating period or cycle lengths on the basis of sparse and unevenly-spaced data.

11.6   The perception of linearity

A linear least-squares fit is a curve fit that is linear in all coefficients: a relationship of the form $y = a + bx$ (where y and x are the data vectors, a is the intercept and b the slope). A non-linear least-squares fit relates to non-linearity in the coefficients. A second difference between the latter and the former, is that for non-linear fitting, preliminary estimates of the parameters are sometimes needed in order to yield a robust result. If a fitted equation serves to derive a prediction of y from x, the fit is called a *regression of y on x*, but if the fit is meant to derive x from y, it is called a *regression of x on y*.

The linear relationship between two variables is about the simplest kind of relationship, yet it has in practice a surprising degree of complication, see Deeming (1968) who points out some subtleties in the analysis of linear correlation in astronomy. Some graphs suggest a linear relationship between the variables: either weak (because the designer [un]consciously wants to see a linear kinship), or strong (because the graph suggests a firm linear correlation: for example, color index (differential log of irradiance) appears to be a straight line in the reciprocal of wavelength in Figure 19. Some non-linear relationships, however, may be linearized, for example power-law relationships[159] will appear as

---

[158]The eye also tends to recognize circular patterns in semi-random noise.

[159]Power-law distributions occur in nature, for example sizes of cometary nuclei and occurrences of solar flares follow a power law, as well as growth processes – see the discussion on the accruement of citations in Section 5 and in Figure 8 of Paper III.

a straight line in a log–log diagram, see also Section 7.3.6. The following points must be considered when dealing with such apparently straightforward data presentations:

1. the strength of the linearity is quite often expressed by a rather weak numerical parameter, the correlation coefficient r;[160]

2. x and y may not be error-free, see the discussion on Figure 29 in Paper III;

3. the errors on both variables may not be independent;

4. the errors on both variables may not be normally distributed;

5. in regressing y on x, or x on y, you minimize using, respectively, vertical or horizontal distance, a choice that is important in view of the preceding points (see Murtagh 1990);

6. if one of the variables is an observable, it is very well possible that some of the values are of better quality than others, thus weight factors[161] may have to be considered.

Least-squares fitting is very sensitive to outliers, hence one should only apply routines for robust regression. The effect of outliers, moreover, strongly depends on their position in the x-interval: a bad point in mid-range may not harm very much, but outliers near the extremes of the x-range definitely change the slope and intercept of the fit.

An example of a moderate linear correlation is shown by Lemaître's available data in Figure 3: with these data, the linearity is not much more than a trend (albeit with a totally erroneous slope) that, with hindsight, is a true linear relation. Let us now make the exercise for the data shown in Figure 3: a straightforward linear regression yields a slope of 477 ± 105 km s^{-1}Mpc^{-1} for Lemaître's data, and 454 ± 75 km s^{-1}Mpc^{-1} for Hubble's. Which one is the most accurate (see also the discussion on page 103), and which one is the most precise? The latter value agrees, within the errors, with the value given by Hubble (1929), *viz.* 465 ± 50 km s^{-1}Mpc^{-1}, and with a mean error on the slope of two thirds of the fit to Lemaître's data, a result that looks more precise. With some reservations, though:

- the data used in this exercise were not published by Lemaître, but were reconstructed from the literature (by Duerbeck & Seitter 2001);

- Hubble even excluded some data, moreover he obtained a different solution depending on whether data were binned or not (the two lines in his diagram, see Fig. 3);

- the *Hubble constant* is a constant of nature that took almost a century to settle to its presently-known value, and as such one may say that the precision at every attempt was quite reasonable, but that the accuracy of the result of every endeavor at determining the Hubble constant is very strongly time dependent and technology reliant (see Fig. 44).

---

[160]$-1 \le r \le 1$; r is the ratio of the covariance σ_{xy} and the product $\sigma_x\sigma_y$ of the standard deviations. When one variable increases while the other variable decreases, the correlation is negative.

[161]For example, the reciprocal of the variance, or a factor based on the quality of the data (Tuvikene *et al.* 2010).

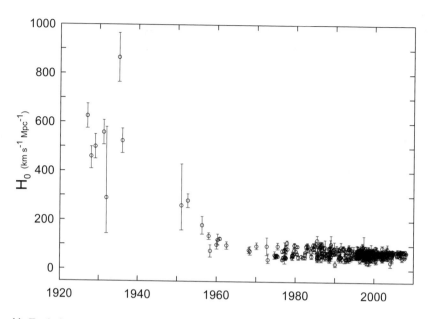

Fig. 44. Evolution of the Hubble constant. Based on data from the *HST Key Project on the Extragalactic Distance Scale*, http://www.noao.edu/staff/shoko/keyproj.html.

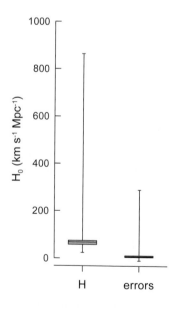

Fig. 45. Box-whisker plot for the Hubble constant and its errors. $\Delta Q = Q_1 - Q_3$ and $k = 1.5$.

	Mean	Median	σ	Q_1	Q_3	$k\Delta Q$
H_0	74.8	68	60.3	25.5	58	75
errors	10.9	8	16.4	10.5	5	12

A straightforward box-whisker plot of the dataset shown in Figure 44, without looking at the graph, may lead the inattentive reader to the conclusion that H_0 is very robustly determined, though it is affected by some outliers in the data and in the associated errors. Figure 45 illustrates this fairly well: data deviating by more than $1.5 \times (Q_1 - Q_3)$ could be flagged as outliers, though in reality they are not. This example clearly shows how dramatic the difference between precision and accuracy can become: consistent observational data have ever led to the seemingly precise but physically inaccurate result that the age of the universe in Hubble's days was much lower than the geological age of our Earth!

A counterexample of a strong suggested linearity that could not stand the test of time is the so-called *Titius–Bode law* of planetary distances.[162] Figure 46 illustrates this relationship for the data that were available in 1766. What is the problem here? Well, the relation

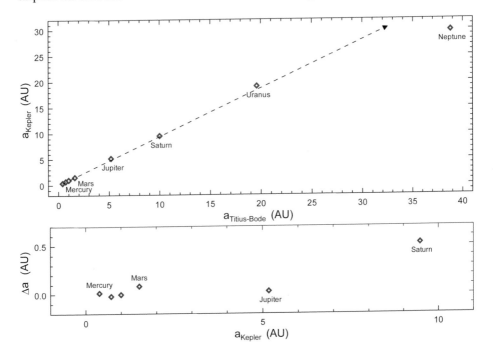

Fig. 46. *Top*: relationship between the semi-major axis a of planetary orbits derived from Kepler's third law (orbital period is the only parameter), and a based on the Titius-Bode law. The dashed line is the least-squares fit for the planets Mercury to Saturn. *Bottom*: differential semi-major axis $\Delta a = a_{\text{Titius–Bode}} - a_{\text{Kepler}}$. The scale ratio of the y axes in both panels is almost 20.

gives a visually perfect fit for the the planets Mercury to Saturn, and this "law" even seemed confirmed when Uranus was discovered by William Herschel in 1781: a linear fit indeed yields a perfect match. But the extrapolation, and the so-called law, completely

---

[162]The semi-major axis a of each planet in the solar system is given by $a_n = 4 + 3 \times 2^n$ ($n = -\infty, 0, 1, 2, 4, 5$), where a is expressed in units of $\frac{1}{10}$ AU and n is the rank number of the planet (starting with Mercury).

breaks down for Neptune. The problem is a problem of perception in the first place, as illustrated in Figure 46 that plots the differential semi-major axis $\Delta a = a_{\text{Titius–Bode}} - a_{\text{Kepler}}$ for the planets that inspired and supported that law: it is evident that the linearity already was a problem from Saturn on. Note that Figure 46 is a comparison of models, not a comparison of observational data.

When purporting linearity, it is much better to support any claims by evaluating the difference between a quantity and its calculated value (or observed *minus* calculated, see the lower panel of Fig. 46), than to graph the full dynamic range of both involved variables (top panel in Fig. 46). The rule to always analyze residuals (see also the discussion on page 79) is especially true for ephemeris determinations of the form Epoch time $= T_0 + EP$, where T_0, E and P, respectively, are the time of an initial epoch, the cycle number and the period: especially when P is very small, E can become extremely large (see Fig. 42), and an E versus time plot will never reveal a cycle-count error in E, whereas a differential plot of observed *minus* calculated time would immediately reveal such errors. This leads to an often poorly understood truism[163] that in such regressions the independent variable E **is errorless, unless it is wrong** – where the first instance of "error" refers to statistical error, and the last meaning refers to error as mistake.

It often happens that the data suggest correlation or linearity, although there is no reason to assume causality. There are, in effect, two questions to be answered, *viz.*,

1. does the correlation really represent a causative case, and
2. if yes, which variable is the cause?

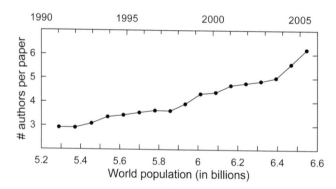

Fig. 47. An example of a nonsense correlation: average number of authors per paper in the non-Letters Sections of A&A and ApJ combined (data shown in Fig. 4 of Paper III) *versus* world population.

The literature is full of so-called *nonsense correlations*, *i.e.*, good-looking correlations that are ascribed to apparently sound causative associations (like total sunspot area and corn price). As an illustrative example, Figure 47 shows the $r = 0.95$ correspondence between the average number of authors per paper in the non-Letters Sections of A&A and ApJ combined (data from Fig. 4 of Paper III) *versus* growth of world population (taken from the *International Data Base* of the U.S. Census Bureau[164]). Even when maintaining

---

[163]A truism is a claim that is so obvious or self-evident as to be hardly worth mentioning, except as a reminder or as a rhetorical or literary device (Wikipedia).

[164]Population Division http://www.census.gov/ipc/www/idb/worldpop.php

that the number of scientists augments when the world population increases, and thus that more people will be expected to share the work, the effect of the increase has a totally different reason.[165] Nonsense correlation is not the same as *spurious correlation*: a relationship suggested by the occasional arrangement of a very small number of data (see Fig. 29 in Paper III).

A special case of nonsense correlation is embedded in the procedure involving *decorrelation*. This is the practice of correcting observational data by so-called "σ-optimization", *i.e.*, by best-fitting an equation that combines, for example, air mass, position coordinates on the detector, seeing, sky reading, etc. The coefficients of the fit then provide minute corrections to the results. If you wish to carry out such kind of data fix, then you must verify that:[166]

1) there is an empirical ground for including a particular observable;

2) the improvement, after decorrelation, is really significant;

3) you are sure that the data statistics is not dominated by outliers;

4) the control data (for example the light curves of the constant stars) still behave properly after the decorrelation procedure, and

5) the resulting parameters behave coherently over time, *i.e.*, that the assumption of correlation is valid for **all** data to which this procedure is applied.

11.7   Curve fitting and overshoots

Fitting a smooth curve through the data is done to

1) summarize a large quantity of data;

2) compare sets of data on the basis of a small number of parameters;

3) select a theoretical model;

4) smooth a noisy signal, thus requiring the curve to pass close to (or within measurement error) through each measured point;

5) allow interpolation, thus requiring the curve to pass *exactly* through each point, or

6) allow interpolation of the smoothed signal (*i.e.*, both previous items combined).

The problem is to fit a smooth curve through a series of (noisy) data points. This can be done with polynomial interpolation, though smooth-curve fitting is mostly done through piecewise cubic *spline fitting*. A spline, literally, is a strip or template of wood (or plastics), or a thin strip of rubber used to draw curves. Figure 48 shows an example of a splinewood. This image is included not merely as an illustration of an antiquated graphics tool, but also to illustrate the effect of illumination on recorded color: this is the only picture in this book captured under outdoor lighting.

Such a tool allowed smooth and continuous connection of a series of points, though with "visual" accuracy, *i.e.*, good-looking unconstrained guesswork, often with excessive

---

[165]The correlation is entirely due to the fact that both variables increase monotonically with time.

[166]For a discussion of these points in context, see Sterken (2010).

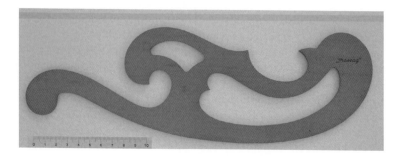

Fig. 48. The splinewood that I used in college in 1964, on a white background, photographed in diffuse daytime sky-illumination. The average *RGB* coordinates of the "white" background are $R = 170.2 \pm 1.5$, $G = 171.6 \pm 1.7$ and $B = 173.8 \pm 1.5$, *i.e.*, a grayish hue. The horizontal band at the top is the background taken from a similar photograph under different sky illumination, yielding average $R = 127.6.2 \pm 1.7$, $G = 157.8 \pm 1.6$ and $B = 203.0 \pm 1.4$, thus a blueish hue (quite close to the lightest × symbol in Fig. 37). The ruler, labeled in 1-cm units, was added to show the scale. This illustration is best viewed in color mode.

curvatures, leading to spurious results. A spline, in our context of computer graphics, is a special function defined, for example, by piecewise polynomial fits (cubic spline or quadratic Bézier[167] curves and surfaces – especially for making vector graphics).

Splines yield smooth lines, but if the derivatives at the ends of the fit interval are not known, the spline fit may turn completely odd. The same happens when gaps in the data series occur. Figure 49 shows an example of a sky brightness measurement as a function of time during the first four hours after twilight, with a fitted spline that shows spurious excursions that will lead any blind interpolation to an erroneous result that is a pure artefact of the applied interpolation method. Such unphysical oscillations – especially when the data are sparse – may be suppressed by adding constraining points nearby the problem areas, or by applying a tension factor that releases the minimum curvature conditions of the fit.[168] Hence, whenever datasets are quantitatively compared on the basis of spline fits, the tension factors should be given along. And, if a smoothing curve is used for analysis, the errors of the fit must be quantified, see Teanby (2007). Spline fitting is very sensitive to outliers, though normal data may also produce outliers in the interpolated dataset, as Figure 49 clearly illustrates. Spline-based extrapolation should be avoided by all means.

11.8   Models and calibrations

The distinction between measurement and model is an aspect that must be preserved at all times. In a not so old paper I read the statement *"The solid line represents a simple sinusoidal fit to the first half of the light curve, and is intended to guide the eye in recognizing*

---

[167] After the French engineer Pierre Bézier, who pioneered – but did not invent – the concept.
[168] The higher the tension factor, the stiffer the fitted curve will be.

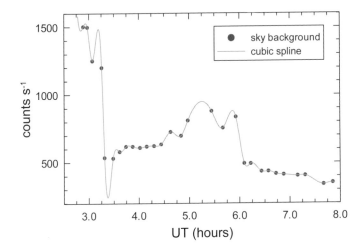

Fig. 49. Example of a cubic spline fit suffering from unphysical oscillations and large overshoots: sky background measurements as a function of time (in detector counts per second).

the asymmetry of the light curve". If a sine curve is added for showing asymmetry only, then the graph will mislead those who do not read the text or the caption. Thus **it is very important to indicate whether a line in a diagram is a fit to the data, or whether it represents a genuine model**. For example, the second-degree polynomial in Figure 8 is a mathematical model that exactly represents the relation between the variables. Likewise, the dashed line in Figure 46 is a fit to data coming from the comparison of two models.

Figures 50 and 51 show another example of differentiation between data and model, in fact between observational data and their calibration in terms of MK spectral type. The photometric color indices of the Geneva photometric system lead to a set of three orthogonal reddening-free parameters X, Y, Z, see Cramer & Maeder (1979), and Cramer (1993) for a definition.[169] Figure 50 shows part of the X, Y plane, where the lines are the calibrations for spectral classes Ia, Iab, Ib, II, III, IV and V. A simple inspection of this diagram gives the impression that deriving spectral class from a set of measured color indices is a straightforward matter that leads to unambiguous and accurate results. Figure 51 shows the same part of the X, Y plane superimposed on the full-range observational X, Y diagram (more than 30 000 stars displayed in Fig. 4 of Cramer 1994). This graph shows that:

1) a substantial part of the X, Y plane is not calibrated – the reason being that the physical role of the indices X and Y changes in the lower part of the diagram, what leads to ambiguity. At more negative Y values, metallicity effects become dominant, and impede photometric spectral classification;

2) X and Y alone can lead to ambivalent results, and

3) discrimination of spectral type very much depends on the X, Y location: especially for stars of types O and early B, it is very hard to derive an accurate spectral type.

This example illustrates, again, the value of graphical support when analyzing data.

---

[169]Not to confuse with the CIE XYZ color coordinates.

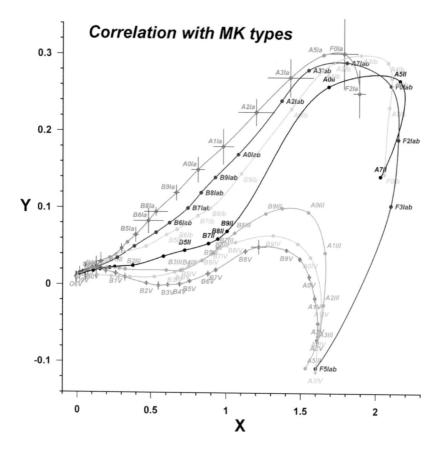

Fig. 50. Schematic diagram of spectral-type calibration in the Geneva photometry X, Y plane. Reproduced from Cramer (2005), with permission.

12   Tables and lists

Tables are used for information that is not suitable for graphical representation, for example observing logs listing exposure times, instrument resolution and scale, observer names, etc. The elements of a Table are often referred to as *cells*.

Tables, just like Figures, are numbered, and their placement follows the same rules as for Figures. It occasionally happens that a Table is presented as a text paragraph, and is not numbered, see the example on page 87.

Lists are used when a small amount of information per item is listed, for example a handful of entries of the form x, y, σ that do not deserve an additional graph. There are two types of lists:

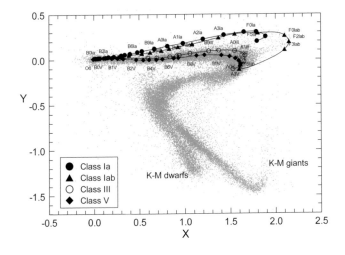

Fig. 51. The X, Y plane shown in Figure 48, superimposed on the full-range observational X, Y diagram taken from Figure 4 of Cramer (1994). Only four spectral calibrations are shown.

1) run-on lists: a handful of items that do not require a specific order, for example in the statement *A&A, AJ, ApJ, and MNRAS are leading astronomy journals*;

2) displayed lists: *e.g.*, bullet lists and numbered lists, with possibly a large number of items. Always prefer neutral marker symbols, including no more than 2 or 3 levels of indentation in 2-column pages, and beware punctuation and capitalization.

Machine-readable Tables are a special case of printed Tables. Some journals offer the possibility to submit data archives for electronic access through an archive or journal website. When importing Tables, you should verify whether

1) missing values are included in between column separators (TAB, OR – or, or ,), and

2) whether there are any empty lines (especially at the end of the file).

Tables often contain lines and rules, and these non-data elements – if needed at all – should follow the same guidelines as in graphics: dense and heavy grids should be avoided at all cost. Horizontal rules are only good for dividing data in topical paragraphs. Always close a Table with a rule (except when there are notes), but a Table with a rule below each line has too much tabular chart junk, and is useless. Tables must be functional, without vertical lines, except when doubling column groups. Color can be a great help for visual clarity, though it does not happen very often (see, as an example, Table 3 in Laurent Cambrésy's paper in Volume 1 of these Proceedings).

The most common typesetting errors in Tables are the column alignment (some authors center every column), the mixing of the math *minus* sign and the dash, the number of significant digits not conforming to the accuracy of the data,[170] and forgetting the leading

---

[170] See Paper I and the discussion of Figure 18.

φ	1874	1875	1876	1877	1878	1879	1880	1881	1882	1883	1884	1885	1886	1887	1888	1889	1890	1891	1892	1893	1894	1895	1896	1897	1898	1899	1900	1901	1902
40																				0.06									
39																													
38																													0.04
37																													0.04
36																												0.02	
35							0.5														0.02						0.1		
34																	0.9			0.1		0.02							
33								0.1									3.6	0.1	0.1	0.02		0.1							
32							0.3	0.1									1.5	0.04	0.2										
31				0.5				0.03	0.1			0.2							0.6	0.2									
30				0.2			1.0	0.1	0.1			0.03					0.2	0.9	3.5	0.2									
29							2.2	0.3	11.2			0.1					0.1	20.5	0.02	0.05						0.01	0.04		
28							5.3	1.0	26.9	0.1							0.5	5.5	25.1	0.3	0.1	0.05					0.1		
27							0.4	4.0	20.8	0.04							0.4	8.9	14.9	3.3	0.5	0.1	0.03					3.3	
26							0.4	23.8	20.0	0.4							0.4	17.8	4.3	2.9	0.1	0.1						11.0	
25							0.1	10.3	8.8	2.1					0.1		0.7	20.0	15.4	6.2	0.9	0.7						2.6	
24	1.2	0.2						16.7	19.0	7.0	0.3				0.3		0.3	19.0	35.1	3.5	3.2	0.1	0.02					2.8	
23	2.4							20.0	20.9	2.4	0.05	0.1			0.1		5.7	25.8	11.7	7.8	5.0	30.7	0.1					0.4	
22	0.1						0.3	36.4	6.6	12.9	0.06	0.1			0.02	10.4	31.4	17	12.6	17.4	25.7							0.4	
21	5.0						21.9	31.0	34.2	28.4	6.3	0.1	0.2	0.04	0.01	8.1	30.2	31.3	14.2	17.7	11.9	0.1	0.04					0.3	
20	5.0						17.1	19.1	82.1	6.1	0.3	0.3	1.6	0.1		12.6	46.5	14.4	14.2	3.8	16.6	2.2	0.04	0.1				0.8	
19	4.2	29.8				0.9	8.7	19.5	8.5	1.3	0.1	1.7	0.1			4.4	38.4	18.1	40.7	11.7	23.4	1.5	0.2					0.04	
18	8.4	4.2					10.2	31.0	9.7	11.8	7.4	2.1	1.3			3.6	32.6	11.2	35.6	21.4	41.5	11.4	0.2	0.3				0.2	
17	1.9	1.6	2.2			0.3	4.5	36.9	33.3	2.6	19.0	8.8	4.8			0.4	22.5	45.3	33.1	27.9	12.8	0.03	0.5					0.1	
16	5.8	3.8	15.3		0.8		8.6	41.2	33.3	3.4	10.1	5.4	2.0	0.2		0.1	27.0	22.5	13.5	25.1	26.3	28.5	1.2						
15	3.7		2.0		2.1		11.9	50.1	25.9	9.9	51.6	16.4	0.7	0.02		0.3	23.2	44.7	34.1	38.7	41.8	10.5	4.8	20.4	0.2		9.7		
14					2.1	0.5	7.6	34	29.7	19.7	51.5	44.3		1.4			17.3	43.8	44.4	27.3	32.3	31.1	8.1	3.4	0.2	0.03	2.8		
13	19.7	3.4	2.8	1.5			15.9	17.2	23.5	27.5	48	13.6	7.3	9.5	0.2		2.4	37.8	23.4	36.4	36.7	20.7	7.5	3.7			2.0		
12	62.1	8.5	4.7	2.1	1.1		2.3	23.6	20.0	30.9	36.4	5.0	5.6	0.7	0.01		35.1	35.2	47.4	44.7	19.8	44.4	12.9		0.7		0.7		
11	19.2	12.6	1.2	1.6	0.7	2.1	4.2	17	38.6	84.8	28.5	15.4	1.3	2.1	0.04		107.4	32.6	19.6	105.3	3.1	12.1	11.6	0.7		1.3	0.03		
10	4.2	34.7	0.9	14.8	2.0		1.2	0.2	17.7	30.4	45.1	44.9	2.7	2.2	0.2		1.1	58.1	25.3	66.3	44.8	3.4	17.9	14.4			3.5	6.2	
9	12.1	13.4	0.4	2.7	2.6			6.5	15.9	30.1	20.3	20.1	3.4	0.7			0.4	37.9	32.7	52.3	25.1	12.5	10.1	1.7	0.1	0.2	10.9	1.3	
8	39.3	16.7	0.4	1.0	2.2			8.9	25.2	42.4	13.9	3.7	0.2			0.05		17.3	31.6	16.8	3.0	4.0	8.7	0.9	0.5	0.8	6.4		
7	28.6	15.2					1.9	7.7	16.5	45.5	21.6	1.9	2.7	0.3	0.9			3.0	1.6	50.4	0.8	1.6	20.9	9.5	2.1		0.2		
6	11.6	1.6			0.6			1.7	2.8	18.3	8.6	1.7	5.8	0.3	1.5			3.2	0.9	19.8	1.9	11.7	18.2	9.4	17.9	0.1			
5	3.4	0.2			9.1			0.1	15.8	16.4	9.6	7.6	0.4	0.7	1.5	0.02	0.04		0.5	7.2	5.9	11.5	13.4	0.1	2.7				
4	2.9	0.2	0.6	0.1	1.5				7.7	15.1	11.1	2.7	3.1	0.3			0.3	0.1	5.0	8.1	8.6	18.7	6.3	0.4	11.0				
3	25.8		0.1	1.6		0.2			7.4	9.4	2.2	0.2	2.0	4.2	0.05		0.1		6.5	3.1	2.3	2.2	10.3	0.1	1.0	1.4			
2	5.8		1.6						1.0	7.4	3.8	4.5	1.1	0.8	1.5				5.0	7.3	4.6	0.5	8.2	1.2	0.2		0.3		
1		0.1	0.1	0.2	0.1				3.7	7.8	3.2	2.5	0.5	1.6	1.0	0.1			0.4	5.0	0.2	0.04	1.2				0.4		
0		0.1							2.4	0.02	0.9	0.4	0.9	0.6	0.03	0.1		0.05	0.5	0.5	0.1	0.02	1.3						
0				0.5					0.9	0.02	0.2	0.7	0.4	4.4	1.0	0.2			2.1	0.05		0.03	3.1						
-1									5.1	0.2	0.7	0.4	3.7	1.0	0.1	0.2			0.2	3.5		0.4	0.2	0.3					
-2			3.4						4.9	0.01	3.5	5.2	10.1	1.1	0.5	0.6		0.1	0.2		0.08	4.0	4.0	0.02	0.03				
-3	3.1	3.6	3.1	0.7							5.7	5.2	1.6	4.7	0.4		0.1		2.7	6.1	2.9	0.05	11.3	0.04	1.3	0.01	0.1		
-4	23.8	0.9	0.1	0.7		0.1			2.1		8.9	10.6	9.4	3.0	0.3			1.6	7.2	8.8	0.3	9.5	0.3	0.1	0.05				
-5	6.0	6.6	8.3				0.1		1.8	1.9	22.1	10.8	3.8	8.1	0.5		0.05	35.3	19.1	2.7	0.4	11.6	0.5	0.6	0.1				
-6	36.3	19.7	4.1	1.2	0.1					10.0	9.9	4.2	27.7	6.4	5.4	0.2		14.1	15.7	14.5	0.2	28.9	3	6.5	0.1				
-7	13.7	2.5	1.9	23.4	0.1				3.9	17	14.1	14.5	11	3.2	8.5	13.8		0.3	36.9	11.5	18.9	2.8	60.9	1.0	8.7	0.3		
-8	21.6	6.7	0.3	0.1				0.02	4.7	28.0	18.1	12.2	7.2	21.4	9.9	11	2.6	5.3	43.3	8.1	9.7	17.9	65.4	9.6	11.4			1.8	
-9	13.7	13.9	4.5					2.2	2.5	52.3	56.9	21.9	12.0	39.1	11.5	7.3	1.2	6.1	41.8	14.2	7.4	17.1	32.1	4.0	1.1	0.4	6.9		
-10	36.7	14.2	19.9				0.4	4.8	5.4	111.5	82.3	17.4	19.6	21.3	5.1	5.9	0.4	7.5	40	14.1	22.8	8.1	26.1	14	16.1	3.4	0.4		
-11	21.6	11.1		0.2			1.0	14.5	19.2	122.9	92.9	25.9	10.7	7.3	0.1	0.05	3.9	39.3	15.1	49.5	30.2	12.3	6.3	21.2	0.6	0.04		
-12	12.5	20.6	1.2	0.2			0.4	8.3	13.3	52.2	55.8	50.8	28.1	4.3	1.1		9.0	64.7	20.7	35	20.9	22.3	14.9	6.0	5.9	0.01		
-13	46.0	3.1	8.8	1.7			7.5	20.5	23.5	42.2	32.8	16	7.2	1.9		3.6	20.7	51.3	81.8	17.5	22.7	11.9	96.5	3.8	8.0		
-14	27.2	18.4	12.9	6.1			4.6	14.1	28.5	59.9	40.1	10.2	1.0	4.0		0.9	19.4	48.2	81.2	31	10.8	13.1	49.2	15.1	2.0		
-15	21.2	4.9	6.2	2.4			3.7	7.8	38.1	53.3	16.7	1.4	0.8	1.2		5.0	36.8	35.2	56.9	26.8	20.4	2.9	14.1	7.5	0.1		
-16	27.8	0.8					4.8	18.9	16.8	38.3	29.7	44.2	28.8	0.9		12.8	17.1	25.7	118	32.4	53.6	3.1	3.4	4.9		0.01	
-17	14.0	1.2	0.6					6.1	6.1.9	19.9	43.5	9.7	0.5	0.1		6.7	34.1	46.4	31.5	39.6	34.7	4.7	1.4	0.1		
-18	41.0						0.7	12.3	16	8.5	28.5	12.5	19	0.9		4.4	5.5	66.7	27.8	24.3	12.1					
-19	6.8			0.5			0.2	18.4	30.0	25.2	15.7	7.2	17.6	14.7	19.4		2.1	5.4	34.1	74.9	38.6	21.8	10.5	14.5	0.2	0.04	0.5	0.1	0.2
-20	7.3						5.0	17.4	31.2	29.2	3.9	4.2	12.9	1.7	0.9	3.7	13	23.1	44.2	43	46.9	8.9	12.7	8.9	0.2		2.0	0.04	
-21							2.1	3.5	5.0	21.1	6.1	3.0	1.1	0.8	0.01	11	4.2	32.7	26.6	62.9	19.6	33	1.2	0.03			1.8	
-22							2.5	7.7	1.6	44.2	1.7	3.0	11.9	0.8		3.8	3.8	6.2	18.0	37.5	4.6	3.4	1.9	29.8			2.1	10.0
-23							11.2	3.3	4.1	37.0	31.3	2.1	1.0	0.2		0.7	1.5	10.7	36.8	5.1	1.0	7.8				0.02		
-24							0.5	6.2	1.5	10.4	23.8	16.2	0.6	0.06		0.4	7.5	9.4	5.2	36.4	9.5	0.2	9.3				0.03	0.04
-25							1.7	7.7	13.9	0.9	4.7	6.5	0.1			4.1	5.2	5.4	10.8	3.8	0.9	2.5					
-26							3.0	1.5	4.6	0.5	17.5	3.7		0.04		1.5	2.4	3.5	3.2	11.2	10.8	28.9	1.2				0.01	
-27							1.8	0.5	5.2	8.5	3.0	2.0		0.01		0.03	1.4	2.2	70.6	13.2	6.4	0.6		0.2			0.01	
-28							0.04	0.9	4.1	40.1	3.8	0.5				0.02	0.9	0.9	40.9	5.1	2.2	0.02					
-29								5.1	2.8	38.2	2.9						0.3	0.9	17.1	3.0	3.2						
-30								0.2	2.4	27.8							0.1	0.3	29.2	1.5	2.1	0.01					
-31								1.5	2.0								0.1	0.2	15.8	0.5	34.8						
-32								0.05									0.1	0.1	0.4	0.1	9.7						
-33								1.6									0.3	0.2	0.3	0.4		0.01				0.1	
-34								0.2	1.2								0.1	0.1	0.01	0.6							
-35								1.6	2.5	1.3								0.02									
-36								0.7																			0.01	
-37								1.8																				
-38								0.6										0.02									0.01	
-40																0.03		0.02									

Fig. 52. Data from Maunder's "anonymous" 1903 paper (reformatted through typesetting). The leftmost column lists heliocentric latitude φ, the column headers are years. Compare to Figure 11.

zero before a decimal (*i.e.*, one should write 0.1000, not .1000 nor 0.1). Another nuisance encountered in Tables, but also in Figure labels and legends, is the mixing of the decimal dot and the decimal comma. Figure 52 is a reproduction of the most important Table of Maunder's "anonymous" 1903 paper, and illustrates the omission of 0-values, the alignment on the decimal dot, and the inconsistent use of the number of decimals (one decimal, except when the data entry is smaller than 0.1). This Table, however, is one of the most beautiful illustrations that a Table is also a graphical element: indeed, the butterfly pattern jumps out without even the need for a plot!

13 What to avoid at all price

The following list summarizes the most common conceptual errors and mistakes encountered in graphics. All these examples lead to unwanted noise in the communication system as illustrated in Figure 2.

- using software that you have not mastered, and letting its wizard take over;

- referring to Figures and Tables that are not included in the text;
- including Figures and Tables to which you do not refer in the text;
- placing Figures and Tables out of order, or too far away from the discussion;
- thin lines;
- unlabeled axes;
- too small text labels;
- using a medley[171] of fonts;
- errors in textual elements, to be avoided by copy & paste instead of retyping;
- wasting valuable space, especially white space around Figures;
- poor contrast with background, resulting in invisible labels;
- mixing 1- and 2-column graphics on the same page;
- destroying artwork by sloppy digitization;
- undocumented image enhancement that manipulates the image towards an aesthetically pleasing result at the cost of data fidelity;
- bad multipliers on axes;
- inconsistent design of graphics in a multi-authored paper;
- using all capitals in axis titles and legends;
- omission of units of measure;
- ticks interfering with the data;
- mixing decimal dots and commas in graph labels and in tabular entries;
- fake perspectives;
- omission of axes;
- using color only for data separation (a fatal error in research-grant applications);
- too dense axis labels;
- varying zoom percentages;
- needless resampling of images to fit the size planned for print;
- image rotations that involve resampling;
- 3-D graphics where the extra dimension is not needed, and
- bending the rules of statistics to prove your point.

---

[171] A medley is a potpourri or collection of sorts of everything.

14 Summary

Before submitting your paper to a journal or to an Editor, the PDF must be 100% portable. Portability can be checked with a so-called PDF *preflight*, the process of detecting problems that may impede error-free printing. ACROBAT, unfortunately, provides full preflight functionality only for its high-end professional products. Preflight output may easily amount to more than a hundred pages for a book like this one. But some straightforward controls can be easily performed:

- check the color pages, and make sure that the color tones and resolutions are OK;

- verify that color illustrations can be read by everyone;

- examine the text proof;

- review the entire manuscript page by page to the end;

- check page numbering to make sure no pages are missing;

- pay special attention to each image: check for shading, missing text and symbols, extra artifacts in the image, background, etc., and

- carefully look at each page.

Communication is the act of successfully conveying information to another reader, but also to a referee, to the Editor, and to the Publisher and Printer. Getting acquainted with the technical aspects and the design facets of artwork is a slow, long and tedious process. But reading graphs also needs training, because the viewer tends to see in them mainly those things which she or he knows in the first place. As McLuhan (1962) puts it

> MY POINT IS THAT I THINK WE'VE GOT TO BE VERY WARY OF PICTURES; THEY CAN BE INTERPRETED IN THE LIGHT OF YOUR EXPERIENCE.

McLuhan, furthermore, explains why non-literate societies cannot see films or photos without much training, and calls for some sort of process of education in reading communication media.[172] Education in reading, and training in making good graphics, can make ideas easier to understand. A poorly designed graph, to the contrary, may render the impression that the underlying science is bad: **a graph can never turn bad science into good science, nor upgrade bad theories into good ones**. Tukey (1977) observed that

> THE GREATEST VALUE OF A PICTURE IS WHEN IT FORCES US TO NOTICE WHAT WE NEVER EXPECTED TO SEE.

Paper III, in its Sections on bibliometrics, includes several graphs that forced me to notice what I never had expected to see.

---

[172]McLuhan (1964) describes medium as "any extension of ourselves".

BECK'S FUTURES 2006
CAN'T WAIT FOR TOMORROW*

INSTITUTE OF CONTEMPORARY
ARTS, LONDON

BLOOD 'N' FEATHERS
PABLO BRONSTEIN
STEFAN BRÜGGEMANN
RICHARD HUGHES
FLÁVIA MÜLLER MEDEIROS
SEB PATANE
OLIVIA PLENDER
SIMON POPPER
JAMIE SHOVLIN
DANIEL SINSEL
MATT STOKES
SUE TOMPKINS
BEDWYR WILLIAMS

*Stefan Brüggemann, Title #287

BECK'S FUTURES 2006
ICE CREAM REVOLUTION*

CCA: CENTRE FOR CONTEMPORARY
ARTS, GLASGOW

BLOOD 'N' FEATHERS
PABLO BRONSTEIN
STEFAN BRÜGGEMANN
RICHARD HUGHES
FLÁVIA MÜLLER MEDEIROS
SEB PATANE
OLIVIA PLENDER
SIMON POPPER
JAMIE SHOVLIN
DANIEL SINSEL
MATT STOKES
SUE TOMPKINS
BEDWYR WILLIAMS

*Stefan Brüggemann, Title #477

BECK'S FUTURES 2006
JOKES, GIRLS AND WOOD*

ARNOLFINI, BRISTOL

BLOOD 'N' FEATHERS
PABLO BRONSTEIN
STEFAN BRÜGGEMANN
RICHARD HUGHES
FLÁVIA MÜLLER MEDEIROS
SEB PATANE
OLIVIA PLENDER
SIMON POPPER
JAMIE SHOVLIN
DANIEL SINSEL
MATT STOKES
SUE TOMPKINS
BEDWYR WILLIAMS

*Stefan Brüggemann, Title #146

SPONSOR'S FOREWORD

This year marks 21 years of Beck's involvement in the UK contemporary art world; the jewel in the crown of which is Beck's Futures.

Conceived in collaboration with the ICA, Beck's Futures allows us to support and exhibit emerging UK artists at a critical point in their career. This year, to further champion artistic talent in the UK, we have made a number of exciting changes: for the first time the exhibition is taking place simultaneously across three major national arts venues in London, Bristol and Glasgow, we've been fortunate enough to have six of the UK's most innovative and respected artists on the award's judging panel and we are encouraging the debate surrounding contemporary art by introducing a public vote element.

Since 2000 Beck's has benefited from an ongoing and fruitful relationship with the ICA and I would like to thank everyone involved for their continued commitment and hard work in helping to establish Beck's Futures' reputation within the art world.

CHRIS STAGG, MARKETING MANAGER, BECK'S

PREFACE

This year marks a bold step forward for Beck's Futures. For the first time, the exhibition takes place in three cities in the UK at once, opening at the ICA in London followed, in short order, by the CCA in Glasgow and the Arnolfini in Bristol. In revisiting the structure of the show, we aim to establish a larger national canvas on which to display the work of the selected artists.

We are delighted to have built an institutional collaboration that offers an exciting and challenging opportunity for this year's artists. Collectively, and as individual organisations, it is important to us to stimulate public engagement and debate around contemporary art. This new, broader platform — and the fact that the public this year have a single 'block' voice on the judging panel — will, we hope, provide greater opportunity for that engagement.

This catalogue includes the work of all the artists shortlisted for Beck's Futures 2006. For the first time, we have also added details of the longlist nominated artists. In doing so, we wanted to offer a context for the choices involved in the final thirteen chosen artists and to give a sense of the richness and diversity of contemporary art practice in Britain today.

As ever, we rely on the efforts of a number of people to make Beck's Futures happen. We are indebted to this year's panel of judges, who have given generously of their time: Jake and Dinos Chapman, Martin Creed, Cornelia Parker, Yinka Shonibare MBE and Gillian Wearing. We are grateful also to the panel of researchers for Beck's Futures 2006 — John Calcutt, Adam Carr, Stuart Comer, Claire Doherty, Alex Farquharson, Francesco Manacorda, Neil Mulholland and Polly Staple — whose research formed the basis for the selection, as reflected in this catalogue. So many elements of Beck's Futures' design and presentation are the result of the efforts of the designers Kirsty Carter and Emma Thomas. We owe them a special vote of thanks for the vision and commitment they have brought to the project.

It goes without saying that the staff at each of our respective institutions have also worked incredibly hard to make an innovative but challenging structure a reality. Finally, we would like to offer our sincere thanks and congratulations to the artists: without their imagination and commitment to the process, none of this would be possible.

EKOW ESHUN, ARTISTIC DIRECTOR, ICA

FRANCIS MCKEE, ACTING DIRECTOR, CCA

TOM TREVOR, DIRECTOR, ARNOLFINI

CONTENTS

STARDUST
BY JENS HOFFMANN

One day you are a star, the next day you are a black hole. — Woody Allen

The artist as celebrity is not necessarily a new concept and has been with us practically as long as the idea of celebrities has existed. Yet, it is not merely the concept of artists as celebrities that connects the art context with the world of show business and glamour; in fact these worlds are often fatally attracted to each other in a variety of ways. While actors or singers become prominent art collectors, or even sometimes artists, artists also add to the cult of celebrity by using stars as the subjects or subject matter of their works both in a critical form and not.

Over the last decades this trend has intensified and the ever-increasing cult of stardom and fame has infiltrated the art context further contributing to the erosion of the distinction between art and its relationship to the masses, blurring the already unsteady borders between art, commodity and pop culture. In this short introductory text I would like briefly to explore what this may mean in the context of Beck's Futures and indeed in the current contemporary art climate.

As an institution based in London dedicated to the display of national and international emerging art the ICA has throughout its history felt an obligation to present and to support the work of younger artists living and working in the United Kingdom. One format in which this support has manifested itself is the Beck's Futures exhibitions and awards which is now already in its seventh year. Over the years Beck's Futures has enabled over 60 younger and emerging artists to present their work to a wide audience, with the exhibitions bringing in a total of over 250,000 visitors to the ICA.

As a new element in this year's Beck's Futures, in addition to the usual jury of art experts, members of the public are asked to give their vote for the final recipient of the award. The intention of this invitation to participate in the selection is to encourage the audience to engage more with the artists in the exhibition, to reflect on the works, to contemplate the visual arts in general and to give them a voice in the discussion surrounding contemporary art. Behind it stands the idea of opening art to a larger audience with the goal of effectively reaching out to more potential consumers. While the concept definitely makes sense on this basic level of increasing audience figures, the discussion will no doubt not be about the nuances of the various works on view but simply about the same old question of the overall legitimacy of art and its often negative and sensation-driven reception within society at large.

The question that arises here, and one that will probably be left unanswered, is whether contemporary art can actually be something that is truly able to reach out to a wider audience without being populist and without becoming a spectacle? While Tate Modern with its continuously high-quality exhibitions holds the official record for most visited contemporary art museum in the world with more than 4 million visitors a year, one must remain doubtful whether contemporary visual art such as that exhibited in Beck's Futures can in fact communicate with the masses. This exhibition is not a retrospective of a famous, historically significant and accomplished artist but a group exhibition of artists in their late 20s and early 30s, artists who are still in the process of developing and defining their practice. It seems only appropriate that one keeps this fact in mind and gives these artists the chance and possibility to develop and grow, to gain experiences without such immediate public attention.

It might be worthwhile briefly to examine society's relationship to artists and the role they play in the overall composition and structure of society in order to understand this debate. The often hostile and tense public relationship to art is a clichéd product of a love/hate relationship. On the one hand, so the story goes, artists are considered to represent a group of people that apparently do not have to stick to the rules and regulations of normal life. They sleep until the afternoon recovering from their indulgence in sex, drugs and rock 'n' roll. They are oblivious to the ordinary problems and responsibilities of the working masses with 9-to-5 routines. This cliché is as much fascinating as it is repellent but rarely lives up to reality. On the other hand it is also the reason why the public loves the idea of the artist as it represents a form of resistance that society can point to when celebrating its supposed diversity and liberalism in allowing for the existence of such lifestyles and supposed countercultures. That those countercultures have long been appropriated, disarmed and commodified by the mainstream is another issue, but the overall notion of the artist in the minds of many remains as a figure of nonconformity and this is perhaps even more so in the UK where the media often portrays artists as a wild group of eccentrics determined to outrage and shock.

A few years ago (before coming to the UK) I did not know that there could be the possibility of a visual artist being a public figure or a celebrity on a par with singers, actors, and fashion models (or other underachievers mostly famous for their pretty faces, their scandals and infidelities rather than any particular skill). Artists represented to me the opposite: people who wanted, through a critical investigation of our world, to give birth to a reflective and inquisitive understanding of society that was, if not entirely opposed to the entertainment industry, at least critical of it. Many artists of the past century have of course deliberately made use of the role of the artist as a public figure and celebrity, addressing thereby the question of what it is to be an artist in a critical and self-reflexive manner. Jeff Koons, Andy Warhol, Marcel Duchamp, Pablo Picasso or Martin Kippenberger all, in different ways, looked at their own status as artists and the role they were supposed to play within a society fascinated by celebrity, but never blandly accepted public attention. Celebrity was just one of the many roles that they took on or used while retaining a critical distance.

Ironically one of the artists who won Britain's most notorious art prize, the Turner Prize, also made the most pertinent piece about the concept of art prizes and their consequences in regard to celebrity and media attention: Martin Creed's Work No. 227: The lights going on and off (2001), which the press did not understand as a poetic rendering of the quintessential duality of life itself or even just as a humorous take on minimal and conceptual art but simply as a direct provocation that aimed at making fun of the audience. Creed in fact put his finger on the main problem of prize culture and the idea of fame: one day the artist is a star standing in the spotlight and the next day the light is turned off, has moved on somewhere else and the artist is literally left in the dark.

Creed's work is just one in a long series of controversial works shown at the Turner Prize exhibitions provoking headlines over the years. Looking at the history of the Turner Prize it becomes evidently clear that such publicity was always part of the overall objective as it guarantees attention for the museum and its exhibitions and not only attracts debate and more visitors but of course also more sponsors.

It is astonishing that the question "Is it art?" (that voiced most frequently in the context of the Turner Prize especially) still gets asked again and again given that the last decades of art production have made it clear that art is not only about technique, or skill but also about ideas, concepts and increasingly about context. On the other hand it is not surprising at all. The truth seems to be that every journalist writing such headlines is part of a pre-arranged game that helps the papers to fill their pages and sell copies and the art institutions to fill their galleries with visitors. We are ultimately manipulated like a reader of a tabloid newspaper into the belief that the public in fact really cares enough about art to be outraged.

Bringing art to society in such populist formats may backfire and contribute to the opposite of a careful and differentiated appreciation of art. Applying the popular formula of vox populi might ultimately achieve the goal of reaching a wider audience but it may also bring the opposite of what most considered art institutions claim at least to want to accomplish: a complex, enlightening and sometimes even entertaining understanding of art.
I would in effect like to end this text with a plea for an evaluation of the artists and the work in this exhibition that respects the right of a younger generation to experiment, to explore and to do so in relative respectful peace. Critical engagement is to be welcomed, but it must begin from a position of recognition of the seriousness with which most artists engage with their work. In general they are seeking neither celebrity nor notoriety, but often to produce work that allows us to recognise such shallow fixations.

JENS HOFFMANN
DIRECTOR OF ICA EXHIBITIONS

BECK'S FUTURES 2006
PARTICIPATING ARTISTS

BLOOD 'N' FEATHERS
PABLO BRONSTEIN
STEFAN BRÜGGEMANN
RICHARD HUGHES
FLÁVIA MÜLLER MEDEIROS
SEB PATANE
OLIVIA PLENDER
SIMON POPPER
JAMIE SHOVLIN
DANIEL SINSEL
MATT STOKES
SUE TOMPKINS
BEDWYR WILLIAMS

BLOOD 'N' FEATHERS

"A snub-nosed laboratory animal drags away a handbag, while with its other scissor-like hand it operates a Moog synthesizer. The tones cause vertigo. 500 dollars is worth nothing now. A beauty of mist rides away on a bicycle. Her rear end can be seen clearly." – Albert Oehlen, Summary of Contents (www.albert-oehlen.de)

Blood 'n' Feathers was born of a certain entourage, namely the reciprocal poetics of Lucy Stein and Jo Robertson, both graduates of Glasgow School of Art. Blood 'n' Feathers fiercely exemplifies a variety of expanded painting -based practice that has long been corroborated in Germany yet largely passes under the radar in quarters of the UK where poetry is mistaken for sentiment and wherein painting is perpetually cast in the irony age. Regularly exhibiting their solo works concurrently, Stein and Robertson have worked in unison on drawings, paintings, collage and tableaux. They use each others' works as catalysts to counteract, and mise en scène to perform in front of. The connections and interplay between their working practices is apparent both in terms of their fleshily ornamented compositions and unabashed facture and in their highly idiosyncratic symbolic perspectives.

Blood 'n' Feathers is a dualism that suggests an oxymoronic ritualistic activity that binds their praxis, a mixture of the sacred and the glamorous. Like the mess left in the aftermath of a cockfight, their paintings are performative. Stein's paintings feature images of winsome men and women abandoned in treacle-textured, sexually charged landscapes festooned with Black Fucking Bunting (2004). The figures they intimate in their works are a combustible medley of voracious dandies and cock rockers, masculine characteristics that they appropriate for their own cruising adventures. Stein's messy-haired heads compare with Robertson's globular finger-painted faces. Many of their paintings focus on friendship through their own relationship, an interest that spills over into their performance in the form of music, provocations, manifestos, diaries and epistolary dialogues, and in live performances such as Robertson's ode to Stein's recent painting The Kiss (2005).

Luscious and assured in passages Blood 'n' Feathers' paintings are also full of fudged spontaneous strivings and capricious gaffes fuelled by a forthright and laviscious sense of humour Bordering Between Fuck You and Fuck Off (2005). Their allusions swarm around mutual heroines such as Courtney Love, part of a contemporary romantic canon of saffron-smeared cultural bulwarks. "In memory of all those who bravely tried to negotiate the subtle abyss", they both exploit an iconography that is as explicit as it is opulent and quixotic. (NM)

JO ROBERTSON BORN 1979 IN MANCHESTER/LUCY STEIN BORN 1979 IN OXFORD.
LIVE AND WORK IN LONDON/AMSTERDAM.

SELECTED RECENT EXHIBITIONS

2006	Jerwood Art Space, London (as Blood 'n' Feathers)
	Collective Art Gallery, Edinburgh (as Blood 'n' Feathers)
2005	CHRONIC EPOCH — THE MOTHER OF ALL PARTIES, Beaconsfield Gallery,
	London (as Blood 'n' Feathers)
	TEAMING, The Embassy, Edinburgh (as Blood 'n' Feathers)
	CAMPBELL'S SOUP, Glasgow School of Art (Robertson/Stein,
	curated by Neil Mulholland)
	KILLING THE ANGEL IN THE HOUSE, Sloterdijk studio space, Amsterdam (Stein)
	AURORA, Edinburgh (Stein, with Shana Moulton)
	HOLY WORRY, Gimpel Fils, London (Stein)
	7 YOUNG PAINTERS, Bowie Van Valen, Amsterdam (Stein)
2004	SOFT WET DREAMS, Mary Mary Gallery, Glasgow (Robertson, solo)
	PANORAMA, The Old Gaolhouse, Glasgow (Robertson/Stein)
2003	THE WINKING BEAR, Glasgow School of Art, Glasgow
	(Robertson, curated by Andy Knowles)
	CELLS OUT, The Old Gaolhouse, Glasgow (Robertson)
	CELIA AND JO, Glasgow School of Art, Glasgow (Robertson, with Celia Hempton)
	KEEPING OURSELVES IN AT THE EDGES, The ArtHouse Hotel, Glasgow (Stein, solo)

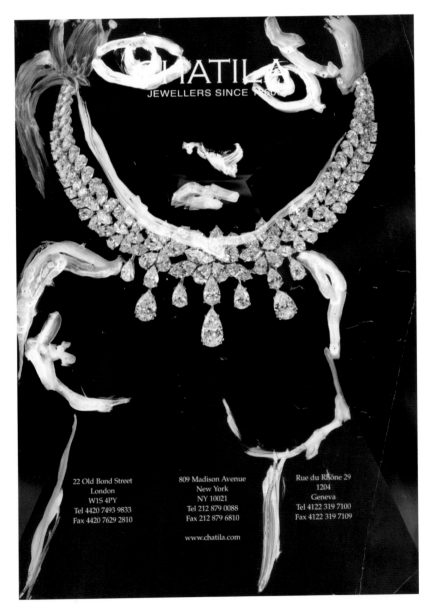

ABOVE: Lucy Stein
PERFECT SYMMETRY (2005)
Ink on paper, A4

OPPOSITE: Jo Robertson
DRAWING (2006)
Tipp-ex on magazine page

FOLLOWING PAGE: Jo Robertson
Assorted sketches (2006)
Various media

EATING OUT
YOUR M
H
E
M

BOYS
Suck

curlie
chips

PRISMS
OF
DESIRE

RACHEL 5 YEARS OLD

PABLO BRONSTEIN

Pablo Bronstein works across a range of media: film, installation, performance and drawing referencing architectural linguistics – from drawing fantasy architectural plans (1780s – 1980s), to constructing physical interventions into galleries and public spaces. Bronstein quotes, references, redraws, reconstructs and reconfigures to establish his own language codes and articulate his own world-view. Although the work may at first appear quietly charming it is in fact bluntly resolute. Ultimately Bronstein is interested in exploring ordering systems and the balance of power in any given situation. All of this is infused with a touch of gender politics, an odd fusion of relational aesthetics and 18th century dandy.

For example a piece Bronstein is working on for the 'Tate Triennale', Plaza Minuette (2005) will turn Tate Britain's Duveen Hall into a controlled piazza through which a group of professional Baroque dancers will proceed in sprezzatura movement. Here Bronstein is interested in how people inhabit public space, the system of manners and bodily comportment of both performer and audience, and how architecture designs how people move. The ink and gouache drawings – 'architectural settings' often in the 'style of' but entirely made up – show the potential of architectural spaces with a certain idea of history in mind. Fountain of Regeneration on the Debris of the Bastille (2005) conjures a time when people built monuments to abstract concepts – 'liberty', 'freedom' – before the democracy of public spaces became a more corporate version.

Monument to the Triumph of the Vernacular (2005) by contrast signals the euphoric post-modern architectural conceits of 1980s run riot. Bronstein focuses on this now tarnished architectural vocabulary particularly that of the 'London School' so vaunted by Charles Jencks, including those such as Michael Graves, James Stirling, Terry Farrell and CZWG. Here a confident post-modern aesthetic – colour, curves, classicism – mixed motifs, styles, historical quotation in a vibrant and carnivalesque post-modern form. Bastardised versions of a po-mo strand of 'new classicism' however quickly became the architectural neo-conservative style of choice, well suited to the dubious optimism and rampant expansion of free market capitalism. Now, subordinate to a super-modern Y2K architectural aesthetic of hi-tech fetishism played out in glass and steel, it all just seems rather embarrassing.

Bronstein is perversely attracted to the pretentiousness, the conservatism and lack of formal integrity of the 1980s, a period when architecture and design sought deliberately to draw attention to itself: buildings as flamboyant poseurs. Combined with a predilection for 18th century romantic vistas and enlightenment planning, through historical juxtaposition and comparison Bronstein acknowledges the arbiters of taste and highlights the despotic nature of architectural creation, and his own. Perhaps this explains Bronstein's attraction to Horace Walpole — by all accounts a total 'show-off', a passionate Gothic revivalist, pioneer of mediavalizing architecture and creator of the romantic architectural pastiche that is Strawberry Hill in Twickenham, London — whose complete works of fiction long now out of print Bronstein has embarked on self-publishing with fancy book jackets designed by himself.

In many cases the institution defines the working space for Pablo; for Beck's Futures 2006 he has proposed a series of architectural interventions. A succession of doorways — finished with cool post-modern curvatures — are set into specially created exhibition display walls. The piece highlights the architectural and stylistic conventions that determine the use of space and at once adds a tension to any straightforward gallery presentation, activating the space for the audience, shifting perspective and behaviour. (PS)

BORN 1977 IN BUENOS AIRES, ARGENTINA. LIVES AND WORKS IN LONDON.

SELECTED RECENT EXHIBITIONS

2006	TATE TRIENNALE, Tate Britain, London (curated by Beatrix Ruf)
	Herald St, London (solo)
	SCENE ONE, Mary Boone Gallery, New York (curated by Jose Freire)
2005	EMBLEMATIC DISPLAY, curated by Catherine Wood for LONDON IN SIX
	EASY STEPS, Institute of Contemporary Arts, London
	HERALD ST PRESENTS PABLO BRONSTEIN, CARY KWOK & DJORDJE OZBOLT,
	Liste, Basel, Switzerland
	MODIFIED UNIFORMS, Blum & Poe, Los Angeles
	SUMMER OF LOVE: ART OF THE PSYCHEDELIC ERA, Tate Liverpool
	(curated by Eli Sudbrack)
	GROUP SHOW, Whitechapel Project Space, London
2004	CURB YOUR ENTHUSIASM, Millers Terrace, London
	IN THE PALACE AT 4AM, Alison Jacques Gallery, London
	PUBLISH AND BE DAMNED, Cubitt Gallery, London
	(curated by Emily Pethick and Kit Hammonds)
2003	BOOTLEG, Spitalfields Market, London

Pablo Bronstein
TRIUMPHAL ARCH IN THE STYLE
OF JUVARRA & LEON KRIER (2005)
Ink and gouache on paper in artist's frame, 36 x 27cm

Pablo Bronstein
STAGE DESIGN IN THE STYLE
OF GIUSEPPE BIBIENA (2005)
Ink on paper in artist's frame, ca. 138 x 48cm

Pablo Bronstein
IDEALISED PERSPECTIVE IN THE STYLE
OF JUVARRA & MICHAEL GRAVES (2005)
Ink and gouache on paper in artist's frame
39.5 x 50cm

OPPOSITE: Pablo Bronstein
LARGE BUILDING WITH COURTYARD (2005)
Ink on paper in artist's frame
44.5 x 50.5cm

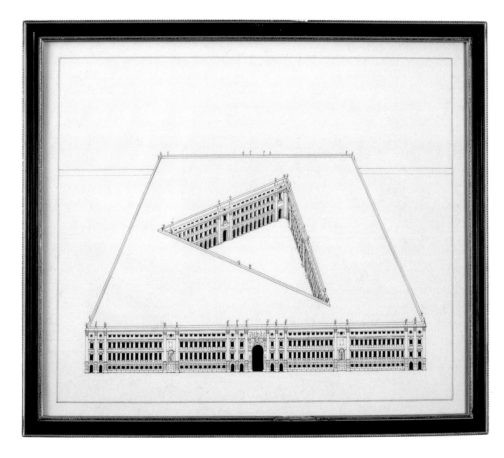

STEFAN BRÜGGEMANN

In one small given moment, how can language both confuse and entertain, or enlighten and critique, narrate and withhold, all at the same time? These questions provide the foundation for the work of Stefan Brüggemann, married to an awareness that any work of art carries within it a history of preconditions.

Often creating neon and text pieces, the established voices of Brüggemann's conceptual forbears – Joseph Kosuth, On Kawara, Lawrence Weiner and Robert Barry – are frequently drawn upon in a game of references that oscillates between the tautological tradition of '60s American conceptualism and wider contemporary concerns. But while the radical investigations posed by Conceptual Art stemmed from a desire to challenge the so-called unquestionable foundations on which art is built, Brüggemann's innovation lies in assuming the position of selector; throughout his work he re-presents and re-forms in order to propagate a new, distinctive voice.

Though seemingly carefully planned, Brüggemann's work often reflects a process of thoughts flowing directly onto paper that are subsequently monumentalized, often as neon or vinyl texts. The neon work Sometimes I Think Sometimes I Don't (2001) exemplifies the strategy of art-as-idea. Which moment within development of 'thought' can be classified as work? And which cannot? Thought, as a product – or rather, the branding of conceptual strategies – appears again in Nothing Boxes (2003), which seems to extend Robert Barry's dictum: "Nothing seems to me the most potent thing in the world."

The context of the exhibition plays a special part in Brüggemann's work – especially for his neon signs, for which the occasion of a show marks a conclusion to the process of making the work. But a settled context is not always apparent; no particular setting could be appropriate for the text work (This Is Not Supposed To Be Here) (2001), and an ambivalence toward the spectator-work-context relationship has been carried forward in a recent work (This Must Be The Place) (2003) for which 'no place' seems to be the 'best place'. With his neons, as with other work, a strategy of endless repetition and reproduction reflects a desire to transform instinctive thoughts into other forms: as one work proclaims, even Thoughts are Products (2001).

In parallel to the artist's neon works, the ongoing work Show Titles (2000 –) demonstrates Brüggemann's fascination with the possibilities that language – and here specifically titles – can pose. The work comprises a list of single words and phrases, usually chosen on impulse, which are offered as possible titles for exhibitions. The list continues to grow, and while the artist sees this as a work in and of itself, he also intends for it to operate as an ongoing archive from which curators and other artists might choose titles for their own exhibitions. "Zebra Crossing", for example, was selected as the title for a survey of current Mexican art at the House of World Cultures in Berlin, 2001.

This process of selection extends directly into his participation in Beck's Futures 2006 where each of the individually numbered titles is to be presented in vinyl text at Show Titles' current form of more than 700. However, Brüggemann's contribution to the exhibition does not end there. He has offered a unique opportunity for three organisers — in London, Glasgow, Bristol — to select one title from the list which will become a title for each of the three exhibitions. In this way, Brüggemann's contribution locates itself both with and beyond the exhibition, and raises a question about selection that is such a key element of his practice. Paradoxically, it may also challenge predetermined qualities inherent in the exhibition: the idea of the competition always hitherto only known as "Beck's Futures 2006", as well as playing with possibilities of repetition across three venues. Here it provides a singular, contextual reading for each exhibition, and demonstrates that slight disturbance or critique can often be camouflaged in a form of complicity. (AC)

BORN 1975 IN MEXICO CITY. LIVES AND WORKS IN LONDON.

SELECTED RECENT EXHIBITIONS

2005   FIVE TEXT PIECES, Blow de la Barra Gallery, London (solo)
        Art Nova, Art Basel Miami Beach (I-20 Gallery, with Joao Onofre
        and Eduardo Sarabia)
        OFF KEY, Kunsthalle Bern, Switzerland (curated by Phillippe Pirotte, cat.)
        THIS PEACEFUL WAR, Tramway, Glasgow (curated by Francis McKee)
        THE FINAL FLOOR SHOW, objectif_exhibitions, Antwerp, Belgium
        (curated by Gerrit Vermeiren)
        (MISUNDERSTANDINGS), Galeria de Arte Mexicano, Mexico City (cat.)
2004   STEFAN BRÜGGEMANN, I-20 Gallery, New York (solo)
        SOLO LOS PERSONAJES CAMBIAN, MARCO, Monterrey, Mexico
        (curated by Patrick Charpenel and Magali Arriola, cat.)
        NOTHINGNESS, Galerie Eugen Lendl, Palais Wildenstein, Austria
        (curated by Neil Robert Wenman)
        WHITE NOISE, The Gallery at REDCAT, Los Angeles, California
        (curated by Clara Kim)
        LA COLMENA, La Coleccion Jumex, Mexico City
        (curated by Guillermo Santamarina)
2003   STEFAN BRÜGGEMANN, Galeria de arte Mexicano, Mexico City
2002   CAPITALISM AND SCHIZOPHRENIA, off-site project, ICA, London
        AGAINST INTERNATIONAL STANDARDS, 24/7 Gallery, London
        STEFAN BRÜGGEMANN, Galeria de Arte Mexicano, Mexico City
2001   THIS IS NOT SUPPOSED TO BE HERE, Museum of Installation, London

Stefan Brüggemann
THOUGHTS ARE PRODUCTS (2001)
White neon

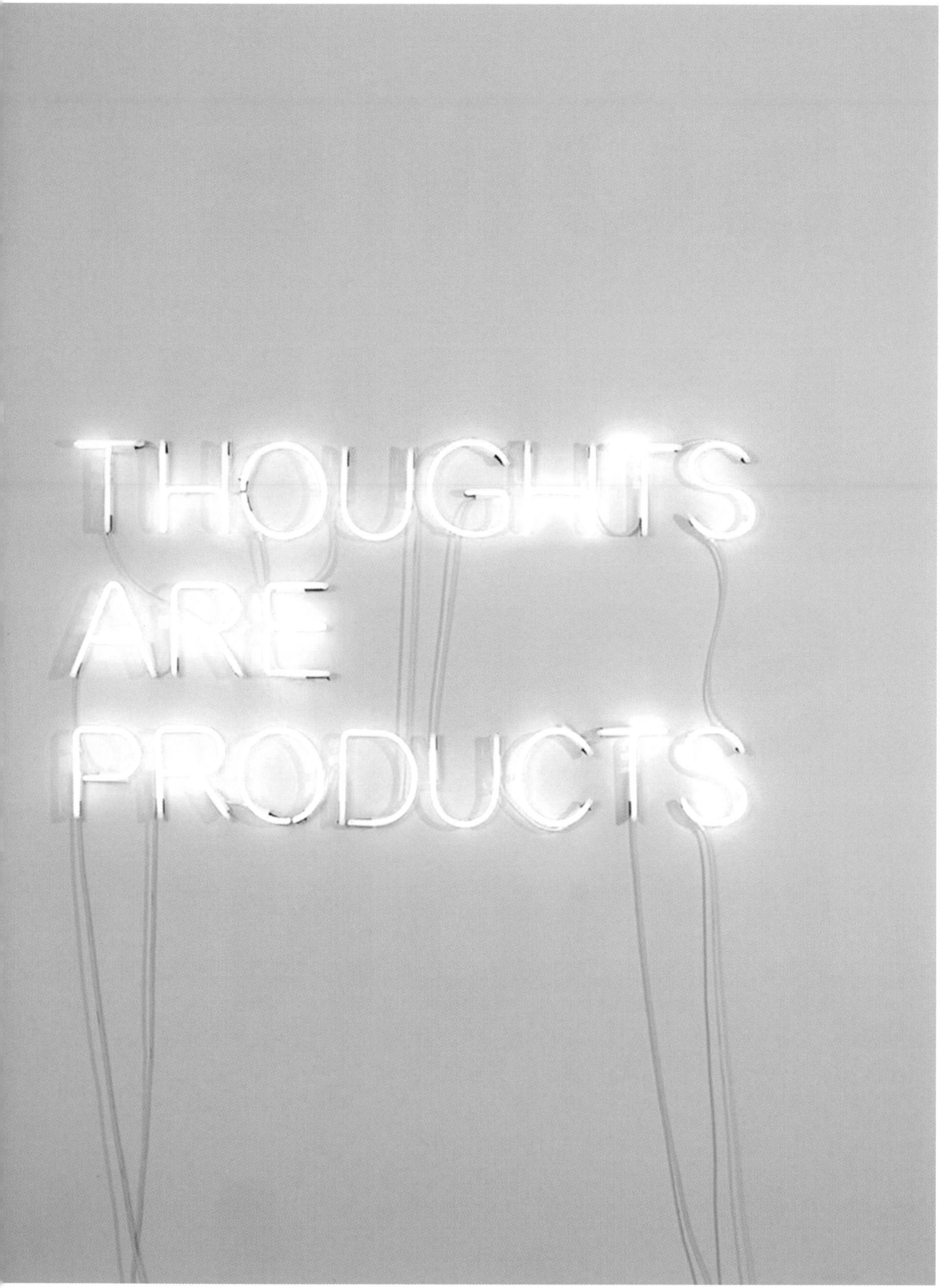
THOUGHTS
ARE
PRODUCTS

SOMET
I THINK
SOMET
I DON'T

MES

MES

Stefan Brüggemann
SOMETIMES I THINK SOMETIMES I DON'T (2001)
Black vinyl text, font Arial Black

RICHARD HUGHES

The work of Richard Hughes aims to put the viewer in an unstable position of belief. His sculptures are similar to three-dimensional trompe l'oeil that, instead of creating the illusion of perspective, create precisely through perspective an illusionary double presence. Shapes such as faces, a tongue, closed or open hands or holding a cigarette and a broken thumb are recreated through the combination of materials incongruous to any faithful effort of simulation. His sculptural vocabulary gathers old mattresses, bicycle tyres, mannequins, burnt fences, bags of rags, rubbish bins and old sofas. Such found detritus is used for the replication of tableaux vivants that have nothing to share with the original materials used to assemble them, rather indulging in being quintessentially artificial.

The artist is interested in creating an imbalance between the objects we see in their concrete existence and three-dimensional dioramas we can piece together in opposition to the first impression of an indecipherable assemblage. A group of legs seemingly involved in group sex recomposes in our gestalt-driven vision as the international 'victory sign' in Love Seat (2005). In Slouching Back (2004), a pile of old duvets and a few pieces of coloured Plexiglas arranged by a wall covered with ripped posters turns into a cloud-crammed sunset if observed again from the correct position. The notion of simulacrum is over-determined and undermined at the same time in Hughes' practice. His work always simulates the shapes of at least two situations, generating an ambiguous competition for truthfulness between materials and perspective's composition. Unlike in figurative painting – trompe l'oeil being the epitome of its illusionism – the material and the medium never disappear in favour of the representational effect.

Although his works all evoke the notion of assisted ready-mades, Hughes painstakingly re-crafts to different degrees of sophistication the objects used for his sculptural arrangements. For Roadsider (2005), what convincingly looks like a urine-filled bottle left on a motorway, is in fact a cast in resin replicating even the small condensation drops on its outside. There is a further mysterious obstinacy to this creation of perfect duplicates of suburban debris, then used to construct compositions simulating something else. The double level of illusion and re-fabrication invests figurative sculpture with a multi-layered remoteness. Hughes' most ambitious replication, collapsing any concept of medium specificity, is Studios5to4 (2003), a copy of a mirror ball whose tiny mirror cubes have been replaced with small photographs replicating the surroundings of the exhibition space where it is installed. Sculpture gets pushed to the utmost degree of imitative fabrication in Hughes' work when even the sculptural raw materials are simulacra. He constantly shifts the status of objects in a process of double illusionism that, like double negatives cancel each other, is almost a concrete attempt to reverse the economy of representation.

In many of the works, Hughes thematically alludes to hippie revivals and to psychedelic pop traditions. He nonetheless articulates them in a distanced way, aiming to include in its homage also a disillusioned credit of its failure – for example the rotten mouth with a red sleeping bag as a tongue that evokes the Rolling Stones' logo in Nan's Ode to Street Hassle (2003). In this sense, Richard Hughes' works combine a fascination for craftiness with a humorous and irreverent re-reading of recent cultural history. His nostalgia never takes itself too seriously, rather it serves as an excuse to obliquely demystify perhaps overly familiar pop references as well as art historical canons and theoretical categories. Every sculptural or representational gesture assumes through its absurdity a conceptual connotation that puts into crisis the notion of imitation from within. (FM)

BORN 1974 IN BIRMINGHAM UK. LIVES AND WORKS IN LONDON.

SELECTED RECENT EXHIBITIONS

2005/06   BRITISH ART SHOW 6, Baltic Centre for Contemporary Art, Newcastle (touring)
           THE PANTAGRUEL SYNDROME, The Turin Triennial Threemuseum Pala Fuksas, Turin
2005   WHAT A DUDE'LL DO, Nils Staerk, Copenhagen (solo)
           RICHARD HUGHES, The Modern Institute, Glasgow (solo)
           THE ADDICTION, Gagosian Gallery, Berlin (curated by Anna-Catharina Gebbers)
           BRIDGE FREEZES BEFORE ROAD, Gladstone Gallery, New York
           ORDERING THE ORDINARY, Timothy Taylor Gallery, London
           DRAWING 200, The Drawing Room/Tannery Arts, London
           HERALD ST & THE MODERN INSTITUTE, Upstairs at Gavin Brown's
           Enterprise, New York
           POST NOTES, ICA, London and Midway Contemporary Art, Minneapolis
           SHOWCASE, City Art Centre, Edinburgh
2004   RICHARD HUGHES, Roma Roma Roma, Rome (solo)
           THE SHELF-LIFE OF MILK, The Showroom, London (solo)
           NO MORE REALITY, Nils Staerk Contemporary Art, Copenhagen
           LIKE BEADS ON AN ABACUS DESIGNED TO CALCULATE INFINITY,
           Rockwell, London
           GRANITE, Whitechapel Project Space, London
           TEALEAF, Cell Project Space, London
2003   EAST INTERNATIONAL 2003, Norwich Gallery, Norwich
2002   SITTING TENANTS, Lotta Hammer, London

Richard Hughes
REAR:
AFTER THE SUMMER OF LIKE (2005)
Couch re-upholstered with hand-dyed fabric,
grease, dust, spraypaint, modelling putty,
wire, enamel paint
FOREGROUND:
THE LAST DRAG (2005)
Cast urethane rubber
Installation view, The Modern Institute, Glasgow

Richard Hughes
FOREGROUND:
STUDIOS5TO4 (2003)
Polystyrene, photographs, chain
BACKGROUND:
LET'S NOT AND SAY WE DID (2003)
Acrylic paint on wall
Installation view,
East International 2003

Richard Hughes
LOVE SEAT (2005)
Polyurethane foam, fibreglass, wood,
clothing, bedding

FLÁVIA MÜLLER MEDEIROS

Using a cross-disciplinary approach spanning the diversity of performance, video and photography, Flávia Müller Medeiros' work occupies an artistic position that explores the way in which language dramatically influences our perception and how this affects representation. Her practice aims principally at examining the established creeds, perception of reality and subjectivities in western society — concerns reflected in a video work Inaugurate (2005) presented in Beck's Futures 2006. Each part of the artist's working practice operates as a process — a complex labyrinth in which works are re-looped, re-visited and often contain other works, in a way suggestive of a Russian Matryoshka doll.

It Depends on the Circumstance and the People (2005), a small yellow book produced by the artist, demonstrates a quite deliberately meandering strategy. The publication catalogues Flávia's artistic output from 1999 to 2004 and at the same time operates as a work in its own right. The artist realises that this chosen format not only provides adequate means for disseminating information about previous work, but can also serve as an occasion to initiate new work that might exist strictly within that format. Performative Concepts (1999 – 2002), a work contained in the publication, indicates this particular condition. As the title suggests, the piece comprises a list of concepts, which constitute the basis of a series of performative works. The artist's aspiration is here directed towards confronting the reader with the imaginative possibilities for the work's form and shape. How would each individual concept directly engage with an unsuspecting public? And what effect could be posed on social groups, or society as a whole? Therefore the outcome of the work is ultimately subjective: it is reliant upon individuals reaching their own conclusions. In parallel, the suggestion of art in the public and the public in art is brought to the fore in a number of works that take to the street, interpolating in daily life. Bringing to mind the pioneering work of Adrian Piper, particularly her Catalysis pieces (1970 –), Flávia's work aims to transpose passive interest into direct participation — a concerted effort to bridge the divisions between art and life.

Although Flávia's work finds affinity with Piper's explorations, her concerns also seem to deviate. For her, it is less about provoking race, xenophobia, and gender issues inherent in society — issues on which Piper's work is founded — and more a proposed form for examining the parameters of socially accepted behaviour. Arguably, today's society increasingly displays impersonal and paranoid characteristics, which Flávia aims to test via acts of direct confrontation and a strategy of charm. One work clearly exemplifying this strategy, You Are Beautiful (2003 – ongoing), could be ephemeral to the point of complete disappearance. But should you meet with the artist, you might be handed a card declaring: "You are beautiful: I made this card to hand out to a beautiful person, especially after eye contact. It intends to acknowledge beauty that is inclusive and subjective. I would like to give as many cards as there are beautiful people in the world." The recipient of the card is left to consider the artist's intent: is her motive sincerity, or deep irony? Is this use of language deceptive? Should we trust this person?

This feeling of ambivalence is taken even further in a video piece Failed, Emerging and Young (2004 – ongoing) where systems of art production become a working principle, and within which the often-unquestionable terminologies used to describe a particular stage in an artist's career listed in the title come under the artist's scrutiny. But rather than an expression of the artist's thoughts, the work invites the views of those who supposedly use these terms the most: curators. Each participant, captured on camera, is asked to articulate their own understanding of what categorises failed, emerging and young artists, and to offer an explanation of how they might use those terms. While seeming to examine the relationship between the position of the curator and that of the artist, it is in fact one aspect of the artist's continual collaboration with others – within and beyond the institution – as means of investigating a broader and more meaningful theme: the multitude of ways in which we use language to give meaning to things, by dissecting meaning hidden within the everyday. (AC)

BORN 1971 IN RIO DE JANEIRO, BRAZIL. LIVES AND WORKS IN LONDON.

SELECTED RECENT EXHIBITIONS & PROJECTS

2005	ABORTED FLIGHTS, Reception Space, London
	ALL AT ONCE, TOGETHER, AT THE SAME TIME, Colony Gallery, Birmingham
	IT DEPENDS ON THE CIRCUMSTANCE AND THE PEOPLE, the artist's space, London (solo, with publication)
	TEMPORARY STUDIO AT THE ICA, I.D.E.A. London, ICA, London
	WISH LIST, M+R Gallery, London
	BOUNDLESS, Stenersen Museum, Oslo
	INTERNATIONAL EXHIBITIONIST, Curzon Soho, London
	STRESS POSITIONS, the artist's space, London
2004	DEFINITIVELY PROVISIONAL, Appendiks, Copenhagen
	XS AN INVITATIONAL EXHIBITION, fa projects, London
	EXAMINING MY OWN PRACTICE, the artist's space, London
2003	INPORT Festival, Tartu, Estonia
	NTH ART, Ols & Co Gallery, London
	MADE IN UK, Arch Gallery, London
	CHOCKERFUCKINGBLOCKED, Jeffrey Charles Gallery, London

Flávia Müller Medeiros
FAILED, EMERGING AND YOUNG (2004—ongoing)
DVD with interactive menu

OPPOSITE: Flávia Müller Medeiros
YOU ARE BEAUTIFUL (2003—ongoing)
Printed cards

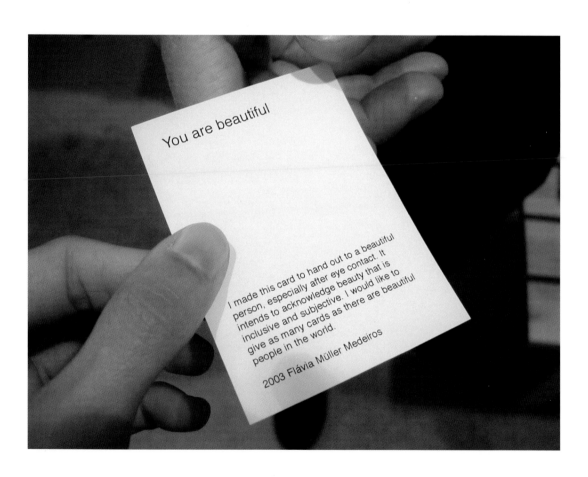

You are beautiful

I made this card to hand out to a beautiful person, especially after eye contact. It intends to acknowledge beauty that is inclusive and subjective. I would like to give as many cards as there are beautiful people in the world.

2003 Flávia Müller Medeiros

Flávia Müller Medeiros
Stills from INAUGURATE (2005)
DVD, 2 mins.

SEB PATANE

A passion for contemporary music and its relation to tribal and primordial roots is the foundation of Seb Patane's practice. This set of interests implicitly translates into his series of drawings over fin de siècle newspapers and Victorian magazine pictures. The images that he selects as a support all share an atmosphere, conjuring up highly staged, slightly repressed scenarios. All of them are produced in the early years of mechanic reproduction when photography began experimenting with fabricated images for public consumption. Onto such controlled and artificial representations Patane adds a secondary layer through a calligraphic gesture. The original images serve as a platform for staging the emergence of inorganic forces.

The resulting modified photographs linger between the drive to obliterate information – in particular human faces, the locus of personality in the picture – and the adding of a second-degree meaning to the image. While concealing, Patane duplicates metaphorically the photographic recording: he seems to be interested in 'correcting' the images that he sources, uncovering their occult unregistered side. In this sense, his drawings are investigations into photography as a paranormal recording tool. The marks that he superimposes onto the pictures seem to be transcriptions that emerge from within the images' own unconscious. The same technique is also expanded three-dimensionally in 93 Dead Sunwheels (2005), a work that transfers the same set of concerns into a more sculptural domain that makes clear reference to music's relation to the primordial and reiterates it in the drawings over two album covers by the esoteric band Death in June.

In the installation Absolute Körperkontrolle (2006), Patane brings to the fore different strands resonating in his practice. He manages to set up a parallel between mountaineering and marathon dancing as similar enactments of bodily rituals. The work consists of a large picture of a mountain expedition from the beginning of the 20th century and a small vintage image of a couple performing in a dance marathon. In the space between, the artist shows a video re-enacting the different phases of a choreographed performance. Against a white background, two male dancers dressed up in mountaineer outfits (designed by Patane) perform contrived exhaustion poses to electronic music the artist himself composed. The dancers' bodies convey the same sense of complete control found in the modified Victorian photographs. The artist orchestrates a similar situation of artificial theatricality, combining various elements of self-discipline hinting at a repressive strategy that ultimately aspires to a liberating finale. The title of the piece clearly sums up this strategy in direct homage to the electronic band DAF (Deutsch Amerikanische Freundschaft), initiators of EBM (electronic body music), whose track Absolute Körperkontrolle is also sampled in Patane's musical composition.

The marriage between dance and mountaineering weaves together intricate links to the occult and hints to culturally acceptable manifestations of archaic impulses. Dance marathons and climbing mountains are both self-imposed competitions of extreme endurance, the former echoing a ritual ceremony of shared perseverance, the latter a more metaphysical challenge opposing human will to natural forces. Patane intertwines these aspects through a substantial although not totally overt reference: the image of the mountain is in fact borrowed from the archive of infamous occultist and mountaineer Aleister Crowley whose legacy can be traced through the entire history of experimental music, from legendary bands such as Throbbing Gristle and Led Zeppelin back to Kenneth Anger's seminal films Inauguration of the Pleasure Dome (1954 – 66) and Thelema Abbey (1955). Patane's hybrid investigations, as well as his fascination for the uncanny in relation to bodily constrictions, expand on the research undertaken in his drawing, mapping out a larger frame of reference that testifies to the artist's passion for the anthropological paradoxes of contemporary culture. (FM)

BORN 1970 IN CATANIA, ITALY. LIVES AND WORKS IN LONDON.

SELECTED RECENT EXHIBITIONS

2006	WHILE INTERWOVEN ECHOES DRIP INTO A HYBRID BODY, migros museum für Gegenwartskunst, Zurich, Switzerland
	Maureen Paley, London (solo)
	Bureau, Manchester (solo)
2005	Prague Biennial 2, Karlin Hall, Prague
2004	THE BLACK ALBUM, Maureen Paley, London
	HA LOO SIN NATION, Lawrence O'Hana Gallery, London (solo)
	CONTRAPOP, Vamiali's, Athens
2003	KICK IT TILL IT BREAKS, VTO Gallery, London
	CHRISTIANE, Shoreditch Electricity Showrooms, London (solo, site-specific sound installation)
	PRETTY LITTLE THINGS, The Ship, London
	BOOTLEG, Spitalfields Market, London
2002	SOUND AND VISION, Royal Festival Hall, London
2001	PENNY DREADFULS, VTO Gallery, London
	UNCHAPERONED, Aroma Project Space, Berlin
2000	SELF SERVICE, 5 Years Gallery, London

PREVIOUS PAGE: Seb Patane
ABSOLUTE KÖRPERKONTROLLE (2006)
Various media
Installation view at migros museum, Zurich

ABOVE: Seb Patane
93 DEAD SUNWHEELS (2005)
Metal music stands, ink on printed paper

BELOW LEFT: Seb Patane
ONE OF THE ALHAMBRA (2005)
Ink on printed paper

BELOW RIGHT: Seb Patane
THE COLOURS AND ESCORT OF THE
HONOURABLE ARTILLERY COMPANY (2005)
Ink on printed paper

ONE OF ~~THE~~ THE ALHAMBRA.

THE GALLANT RESERVE FORCES OF OLD ENGLAND.

THE COLOURS AND ESCORT OF THE HONOURABLE ARTILLERY COMPANY.

OLIVIA PLENDER

An artist and prolific writer, Olivia Plender is enthused by a curiosity regarding paranormal phenomena, suggestion and persuasion. Working across an eclectic array of media her practice helps to recover the shadows of the enlightened modern era. A rich web of associations emerges from her research into magick, parapsychology and spiritualism. The outcome of these investigations ranges from lectures, such as a paper on the Spiritualist National Union given as part of a religious recruitment performance event A Crack in the Cosmic Egg at Cell 77, Edinburgh in 2005, to exhibitions such as The Medium and Daybreak at Castlefield Gallery in Manchester in 2005.

Her mock-horror graphic serial The Masterpiece (2003 –) appropriates the narrative and pictorial tropes of American comic books, 19th century illustrations and art history to produce a sardonic hybrid. The result is a comic pitched between the anti-heroic fantasy of Neil Gaiman's Sandman (1989) and the social realism of Marjane Satrapi's Persepolis: The Story of a Childhood (2003). The Masterpiece tells the story of Nick, a painter living in 1950s bohemian London who plans to impress the Soho elite by making the ultimate painting. While Plender's drawings may lack the confidence of a professional comic book artist, the origins of Nick's 'genius' are nevertheless presented in a style recognisable to anyone familiar with DC and Marvel superhero narratives such as Superman and Spiderman. This customary tale is interrupted in Issue 2 when Nick's lover dreams of an ectoplasmic spectacle of Victorian spirits. The two narratives combined remind us that the romantic stereotype of the artist and newly invented paranormal pseudo-religions commonly intermingled in the late 19th century when many modernist artists readily absorbed mysticism as a radical political gesture and alternative bohemian lifestyle.

Today 'artistic genius' remains a dominant myth, the practices of artists being equated with something akin to mediumistic automism, an unconscious creative act beyond the control of mere mortals, a gift born of a exclusive rapport with the Muses. Despite the obvious fabrication of such narratives and myths they continue to permeate many aspects of our experience of art. In this sense, Plender embraces the subject of magick as a means of reading the exhibitionary complex beyond the bounds of modernism and post-colonialism. In Plender's work the spectres of the dime museum and the freak show coalesce into a broader fascination with sleight of hand and mythologisation both of which, it might be argued, lie at the core of art practice. Plender's work explores forms of spectacle, illusion and spellbinding in creative ways that don't simply serve as predictable demonstrations of power and subliminal control as is so often the case with art that proffers institutional critique. Rather they suggest a complicit and participatory audience is necessary for cultural rites to function. (NM)

BORN 1977 IN LONDON. LIVES AND WORKS IN LONDON.

SELECTED RECENT EXHIBITIONS & PROJECTS

2006	A STELLAR KEY TO THE SUMMERLAND, to be published by Bookworks
	TATE TRIENNALE, Tate Britain, London (cat.)
2005	THE MEDIUM AND DAYBREAK, Castlefield Gallery, Manchester (solo)
	A PUBLIC MEETING TO ADDRESS THE PHENOMENON OF MATERIALIZATION,
	Man in the Holocene, London (solo)
	IX BALTIC TRIENNIAL OF INTERNATIONAL ART, Contemporary Arts Centre (CAC),
	Vilnius and Institute of Contemporary Arts, London (cat.)
	A PRESS CONFERENCE WITH KEN RUSSELL, performance at Camden Arts Centre, London
	PARIS-LONDRES: LE VOYAGE INTERIEUR, Espace Electra, Paris (curated by Alex
	Farquharson and Alexis Vaillant, cat.)
	CONISTON WATER FESTIVAL, Grizedale Arts, Cumbria
2004	SOLO SHOW, VRC at Dundee Contemporary Arts, Dundee
	ROMANTIC DETACHMENT, PS1/MOMA, New York (touring to Chapter Gallery, Wales;
	Q Gallery, Derby; Folly, Lancaster and Grizedale Arts, Cumbria)
	EAST END ACADEMY, Whitechapel Gallery, London
	FOR EXAMPLE, Hayward Gallery, London (performance with Christopher Howell)
	PROJECT #01, curated by Assembly projects, Liverpool Biennial
	PUBLISH AND BE DAMNED, Cubitt Gallery, London
	(touring to The International3, Manchester)
2002	RATIONALIZE, STANDARDIZE, MECHANIZE Plattform Gallery, Berlin, Germany (solo)

THIS AND FOLLOWING PAGES:

Olivia Plender

From THE MASTERPIECE — ISSUE 3, EVIL GENIUS (2004)

Published by Grizedale Arts

Of course he was the outsider against society, but he wasn't *conscious* that he was an outsider. Remember that up until quite recently the outsider didn't have any chance and now it's a great bonus. One is *more* likely to succeed than not.

SIMON POPPER

Simon Popper's photographic and conceptual work elaborates on a refined passion for quotation that he painstakingly weaves together with elegance and exactitude. Every project that the artist undertakes implies some sort of identification with a combination of literary heroes, film auteurs, ideas or art historical references that are then gathered to draw a configuration of hidden correspondences. The elements vary from language-obsessed authors such as Marcel Proust or James Joyce, to humorous and combinatory conceptual artists — Alighiero Boetti in particular but also Douglas Huebler and John Cage — to a complex set of interwoven autobiographical references.

In Beck's Futures 2006 the artist shows Borromean (2006), a new work sketching out intellectual and fortuitous connections emerging from the most expensive book published in the 20th century. Popper has reprinted the first edition of Joyce's Ulysses, identical to the original except for the fact that every word in the text is re-ordered alphabetically. In order to match the first edition of three published formats, the thousand newly-printed books are divided into three stacks reminiscent of Carl Andre's formal arrangements.

In Lacanian theory, the Borromean knot serves to aid visualisation of the ties between the three psychic orders of the Real, the Symbolic and the Imaginary. It is a diagram showing a knot whose three rings are liberated if one of them becomes detached. Lacan also hypothesises a fourth ring holding the fragile bond together: it is the 'sinthome', the signifying articulation of the subject beyond meaning. The 'sinthome' denotes the way in which the unconscious determines the subject's jouissance, which Lacan exemplarily equates to Joyce's writing as a private strategy to avoid psychosis and keep the three orders together.[1]

Popper replicates such ultra-signifying activity as the element compacting his system of literary and theoretical quotations. This is the piece called

Sinthome consisting of a toy train circling on a track around the book sculptures. Its carriages are the faithful reproduction of those belonging to the former transport company Blum & Popper. Allegedly, Joyce transliterated Luis Blum's name, one of the owners of the Trieste based company Blum & Popper, later known as Adriatica Società Anonima di Spedizioni of which the Irish writer was a client, into the surname of Ulysses' protagonist (Leopold Bloom), combining it with the forename of the other owner Leopold Popper.

Consistently, Simon Popper's projects are carefully composed constellations held together by a closed system, a tight set of rules serving to reiterate the work's content on a formal level. The artist plays with a recursive strategy that doubles his references' concerns in the conceptual and practical production of the piece. This strategy aims to create a sealed procedure that assumes symbolic connotations. One such impenetrable process is illustrated in the photograph Things I Might Have Learnt to Ask Him, Can No Longer Be Answered (2002). The artist has taken a real size picture of the image made by his grandfather on the morning before his death, depicting the swimming pool where the event occurred. The single image enclosing a self-referential system of image production represents for Popper almost both an operational model and a poetic strategy.

A similar scheme is at work in Facades (2005), a complete collection of photographs depicting all the London embassies, buildings that serve as self-representations of foreign countries, a fragment of sovereign territory within another nation. This work exemplifies Popper's serial practice in which completion of a given task recursively seals the piece's conceptual architecture. The same mathematical principle of closed entities contained within bigger constituencies is further explained in the title of the work Frontiers Describe What is Beyond as well as What is Enclosed (2003). In this series of blueprints, Popper outlines the borders of all the states that exist within states, in homage to the enigmatic rigour of Boetti's series 12 Forme dal Giugno '67 (1969) consisting of copper etchings of maps outlining the nations at war in June 1967. (FM)

Note
[1] Sinthome is an old French way of writing symptom but also includes the puns 'synth homme' (synthetic man) and 'saint homme' (saint man). Lacan defined it in Jacques Lacan, Le Séminaire XXIII (1975-76), Le Sinthome, Edition du Seuil, 2005 part of which can also be found in Jacques Aubert (ed.) Joyce avec Lacan, Navarin éditeur, 1997.

BORN 1977 IN LONDON. LIVES AND WORKS IN LONDON.

FOLLOWING PAGES:
PANTONE 3305, and Simon Popper
BORROMEAN (2006)
Proposed configuration of 1000 reprinted copies
of the first edition of James Joyce's Ulysses

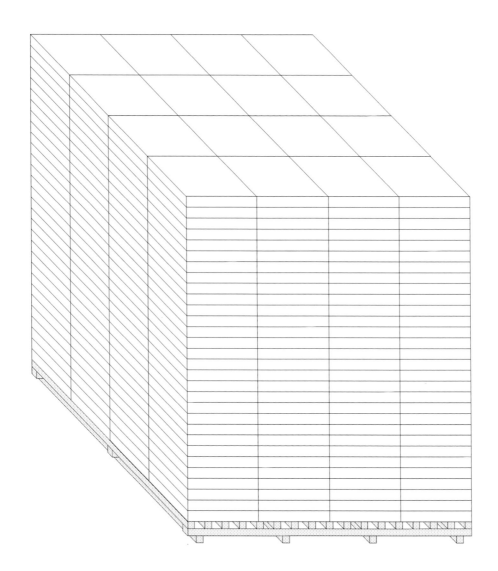

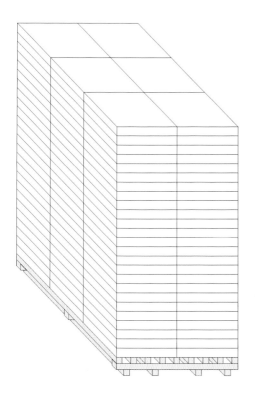
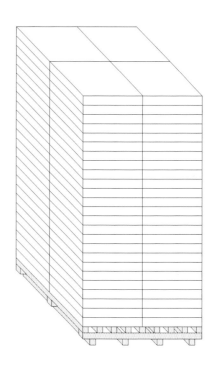

JAMIE SHOVLIN

In Flaubert's famous novel Bouvard and Pécuchet the two protagonists, in the midst of unsuccessfully dedicating themselves to be total-experts of a sequence of disciplines, decide to run a museum in their house and start compulsively collecting dubious archaeological material. Their intention is to make sense of the multiplicity of the world around them by defining precise limits of knowledge, an ineffective attempt to 'divide and conquer' the real. The heterogeneity of their collection represents their success and their failure at the same time. The gathered fragments acquire sense only when an utterly personal eccentric history is projected onto them.

The work of Jamie Shovlin seems to push to an extreme both the necessity and the absurdity of idiosyncratic systems of ordering the world. The artist shares Bouvard and Pécuchet's passion but understands that to be effective such an enterprise needs a demarcated area. Faced with the impossibility of finding such unbreakable boundaries in contemporary knowledge, Shovlin paradoxically reverses the notion of scientific demonstration and gives in to the necessity to create a fictional domain from scratch. Once the artist has selected a topic to investigate, he sets in motion a mechanism that organises, forges and hierarchizes the materials at stake through the construction of a highly subjective mythology. The archive that gathers them represents a collection of evidence, simulated traces. These are leftovers testifying to an existence; for example the Naomi V. Jelish Archive (2001 – 04) collects such diverse records as sketchbooks, letters, photographs with friends and small notes with tea spills.

In his new project, Shovlin has brought together an archive of memorabilia about Lustfaust, supposedly a legendary band from the Berlin experimental music scene of the late '70s. Due to an innovative mode of distribution (you could send an empty cassette to the band and they would send it back with recordings), the biggest part of this fictional collection consists of hand-drawn covers of bootleg cassettes designed by band devotees archived by collector sometimes with attached memorabilia from the fan. Also in the archive there are fantastically designed posters, concert reviews, the fanzine Falke Tränen, and rough samples of the music. But the fictional construction goes beyond the production of material evidence and invades the display strategy, as the work is presented as a real archive open to consultation.

In this fictional architecture, the artist disguises himself as the exhibition's curator, assuming a position of scientific detachment corroborating the whole structure. However, Shovlin is not so much interested in convincing the viewer of the verity of the universe he is forging, but rather aims at seducing us to step into a fictional alternate world, and momentarily suspend our disbelief as we do at the cinema or when reading literature.

In the work Fontana Modern Masters (2004 – 05), Shovlin has set up a mathematical system that reasonably predicts the possible covers of the planned but unprinted books in the series of the same name. Following an accurate measurement formula illustrated in a mandala-like chart, the artist has painted watercolours of the unpublished books, applying analytic exactitude to an entirely hypothetical and futile subject. Failure and absurdity are built-in as conditions that allow the work to come to life. Fontana Modern Masters explores the region of the unrealised as a paradigmatic area to test out an over-determined system of classification. Arbitrary processes such as cover design and editorial decisions about key intellectual figures are treated as if obeying a statistically precise master plan. The role of constructed theories and narratives in the conception of the work exposes, by negation, the mythological nature of any more or less obsessive attempt to give the world a definitive sense by the means of a neat and rational order. Shovlin's art is based on the ambivalence towards these doomed efforts; on the one side we reject it as inauthentic, and on the other end we are compellingly drawn towards its stabilising effect. (FM)

BORN 1978 IN LEICESTER. LIVES AND WORKS IN LONDON.

SELECTED RECENT EXHIBITIONS

Forthcoming	AGGREGATE, ArtSway, Sway, Hampshire; City Gallery, Leicester; Hatton Gallery, Newcastle; Talbot Rice Gallery, Edinburgh (solo)
	A HISTORY OF THE MODERN WORLD, Central Space, London (solo)
2006	IN SEARCH OF PERFECT HARMONY, Art Now, Tate Britain, London (solo)
2005	FONTANA MODERN MASTERS, Riflemaker, London (solo)
	AFTER THE FACT, Tullie House Museum, Carlisle
2004	FOR SOME OTHER CAUSE, IBID Projects, Vilnius, Lithuania (solo)
	GALLEON AND OTHER STORIES, Saatchi Gallery, London
	ARTFUTURES, Contemporary Art Society, City of London School, London
	THIS MUCH IS CERTAIN…, Royal College of Art Galleries
	INSIDE OUT – INVESTIGATING DRAWING, Milton Keynes Gallery
	NAOMI V. JELISH, Riflemaker, London (solo)
2003	NTH ART, Ols & Co. Gallery, London
	BLOOMBERG NEW CONTEMPORARIES 2003, Cornerhouse, Manchester and 14 Wharf Road, London
	PLEASE – TAKE ONE, '39', London
2002	FIVE BOOKS, Hockney Gallery, Royal College of Art (solo)
	ARTLAB22: OVER THE ROAD, Imperial College, London
	SPECTRUM II, Heathcote Arts, Nottingham
	DIVERSION, Arch 295, Camberwell, London

Jamie Shovlin
LUSTFAUST: A FOLK ANTHOLOGY 1976 – 1981 –
GIG POSTER – BRUXELLES NOVEMBER 1979
(2004 – 06)
Collage, ink on paper, 32 x 25cm

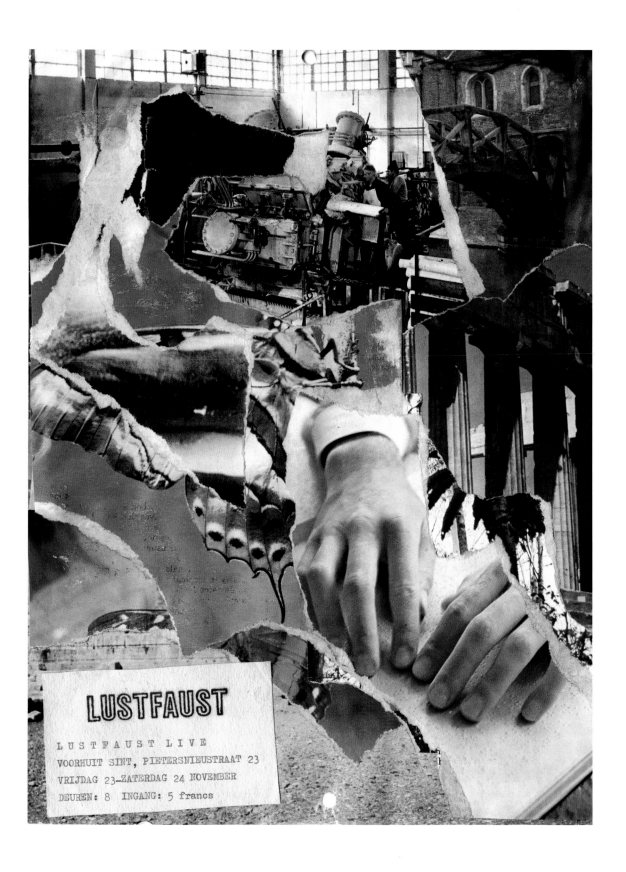

LUSTFAUST

L U S T F A U S T   L I V E
VOORHUIT SINT, PIETERSNIEUSTRAAT 23
VRIJDAG 23–ZATERDAG 24 NOVEMBER
DEUREN: 8   INGANG: 5 francs

A FOLK ANTHOLOGY 1976–1981

LUSTFAUST

CURATED BY JAMIE SHOVLIN

FROM THE ARCHIVES OF
MIKE HARTE & MURRAY WARD

OPPOSITE: Jamie Shovlin
LUSTFAUST: A FOLK ANTHOLOGY 1976 – 1981 (2004 – 06)

Jamie Shovlin
LUSTFAUST: A FOLK ANTHOLOGY 1976 – 1981
– ÜBERBLICKEN∕ÜBERZEUGEN
BY MICK LARKIN (2004 – 06)
Pen, felt-tip, pencil and letraset on paper,
card mount, ink stamps, 19 x 26cm

DANIEL SINSEL

Daniel Sinsel's carefully constructed oil paintings and assemblages derive
from a magical alchemy of neo-classical painting, agonising beauty, surrealism,
and straight-up sexual frisson. Sinsel produces erotically charged talismans
for a private utopia; witty creations that could be devotional icons for a
personal and pagan mythology. Although his images are frequently culled
from the pages of homoerotic magazines, he transforms the pedestrian
desires of pornography into lyrical meditations on a more elusive — if slightly
sinister — variety of longing.

Small in scale, and frequently incorporating evocative materials such as
horsehair and gemstones, eggs, nuts, and precious woods, his art teases the
viewer with its small scale, tactility and portability. Sinsel's cameos convey
the intimacy of gifts or offerings, and the desire to touch, hold, or physically
engage with them threatens to overwhelm the traditional contract between
viewer and painting. His work is a dexterous reassessment of painting's
haptic charge as both image and object. It bridges the world of dreams with
that of physical reality.

Sinsel's imagery conjures a wide variety of historical precedents but dislodges
them from any familiar narrative. One painting on a small linen disk
features the penetrating gaze of a young man. Disembodied and situated
next to a hazelnut and two rosehips, his mysterious countenance recalls a
2nd century funeral portrait of a youth from Roman Egypt. Another work
presents two boys who might have been harvested from an Italian Baroque
ceiling fresco. Their flirtatious glances dart out from a pair of plaster
eggshells joined and suspended by a leather cord. Francisco de Zurbaran

seems a possible influence on the softly-lit folds of a solitary radicchio in one still life, and Sinsel's rendering of a boy pleasuring himself in a bed of grass evokes the taut physiques populating the mid-20th century pastoral paintings of Paul Cadmus and Jared French.

The accumulated sense of déjà vu in Sinsel's work should not imply that his images are retrograde pastiche. Rather, they re-position the familiar and the antique into intimate, esoteric emblems that disturb the coherence in art history's conventional chronologies. They chart the course of a fugitive eroticism through its various historical manifestations, scrambling any logical union between the silent yearnings of the past and the urgent desires of the present.

Many of Sinsel's works seem to function like totems of fertility and longing, but their meticulously crafted seductions are often accompanied by darker, more menacing motifs. Pin pricks, gaping mouths, flayed skin, scars and sharp knives repeatedly materialise on his exquisite surfaces, creating uncanny visual puns about the fragility of skin and surface. Silken ribbons writhe around the heads of ivory-skinned youths, seemingly offering them as gifts, but equally threatening to bind them forever in their luxurious filigree.

The knives, phalluses, eggs, orbs and other objects that populate Sinsel's work form a symbolic order similar to that of Surrealist writers like Georges Bataille, who drew comparisons between eyes and eggs in his writings and sought to violate and mutilate these forms as a means of transgression. For Bataille, the act of looking is informed by seduction, and death is the pure and final demonstration of this eroticism. Sinsel similarly situates his work at the boundary of seduction and pain. His images are spirited memento mori that heighten the tension between the physical and the visual and occupy the present and the past in equal measure. (SC)

BORN 1976 IN MUNICH, GERMANY. LIVES AND WORKS IN LONDON.

SELECTED RECENT EXHIBITIONS

2006	Luisa Strina, São Paulo, Brazil (solo)
2005	LONDON IN ZÜRICH, Hauser & Wirth, Zurich
	Sadie Coles HQ, London (solo)
	THINK & WONDER, WONDER & THINK, Museum of Childhood, London
2004	NOW IS A GOOD TIME, Andrea Rosen Gallery, New York
	YEAR ZERO, Northern Gallery for Contemporary Art, Sunderland
2003	EXPLORING LANDSCAPE: EIGHT VIEWS FROM BRITAIN, Andrea Rosen Gallery, New York
	DIRTY PICTURES, The Approach, London
	STRANGE MESSENGERS, The Breeder Projects, Athens
2002	BLOOMBERG NEW CONTEMPORARIES 2002, Barbican Art Centre, London
	and Static, Liverpool Biennial (cat.)
	APRILFOOLS, Erika Verzutti, São Paulo, Brazil
	SHOWCASE 2, London Institute Gallery, London

Daniel Sinsel
YOUNG MAN WITH RADISH (2005)
Oil on linen, 35.8 × 30.5cm

Daniel Sinsel
UNTITLED (2005)
Oil on linen, 29 x 23.5 x 6cm

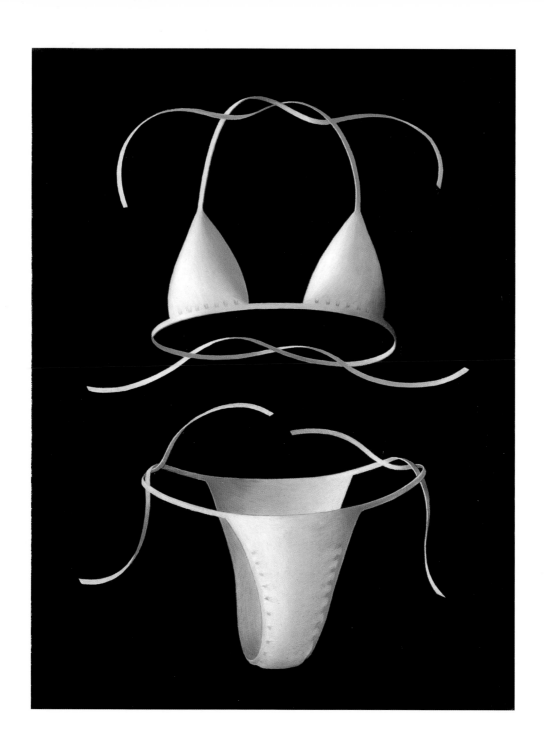

Daniel Sinsel
UNTITLED (2004)
Oil on linen, 41 X 30.5cm

MATT STOKES

Matt Stokes' artistic practice is marked by anthropological enquiry. He is intrigued by informal movements or events that bind people together. Investigating, interviewing and assembling the marks of belonging – the music, the clothes, the memories, the moves, the chat – Stokes explores what occurs when an outsider restages or represents others' experiences.

His early work concerned the history of Out House Promotions and the cave raves, a series of rave events organised in 1991 around the Lake District area of Coniston. His ongoing project Real Arcadia (2003 –) archives and interprets the clandestine experience of those who attended or were affected by the events. By historicising this amorphous and largely invisible collective, Stokes has created a project which pushes the received identity of rave culture beyond the media hype, revealing the impact of its lost ideals on its survivors. Projects in 2001 – 02 involved interviewing residents of Sunderland in the back of a limousine, broadcasting an all-time-top-40 countdown of music tracks for the residents of a tower-block and chartering a steamboat on Kielder Water in Northumberland.

His most recent projects, commissioned by Dundee Contemporary Arts, involved the investigation of the local music scene. Long After Tonight (2005), was filmed on Super 16mm, and documents Stokes' restaging of a Northern Soul dance event at a Gothic Revivalist Church. Black American Soul music first became popular in the North of England in the mid-sixties. By the seventies 'Funk' had taken over in the South, but the North remained loyal to R&B. The type of venue for Northern Soul differed from the intimate clubs of the South. The North continued to utilise much larger venues such as abandoned Victorian dance halls, seaside piers and churches like St. Salvador's. Stokes' film perfectly captures the movement of the spins, flips and drops, lending the event an epic quality.

St. Salvador's Church also hosted Stokes' Northern Soul pipe organ recital, one of three recitals to form Stokes' other Dundee project Sacred Selections

(2005). Representatives from underground music cultures strongly associated with Dundee – Northern Soul, Happy Hardcore and Black Metal – selected indicative pieces which were then transcribed for the churches' elaborate pipe organs by The Royal College of Organists. One would expect the combination of modern music and church organs to result in somewhat incongruous renditions, but the recordings reveal something all the more surprising. A Sy and Demo hardcore track becomes an uplifting hymn, whilst The Hordes of Nebulah is transformed into an operatic score.

Stokes' astute response to his subjects of enquiry is to acknowledge the distance between him, them, and us – the audience. He employs strategies such as performance or film to explore re-enactment as a way of representing a collective experience of the recent past. In doing so, he avoids the pitfalls of documentary, and subverts his own authorial voice. And by involving other specialists, participants and members of the collectives, he acknowledges the multiple and conflicting experiences that contribute to their sense of belonging. (CD)

BORN 1973 IN PENZANCE, CORNWALL. LIVES AND WORKS IN NEWCASTLE-UPON-TYNE.

SELECTED RECENT EXHIBITIONS & PERFORMANCES

Forthcoming	Collective Gallery, Edinburgh (solo exhibition and commission)
	EAST INTERNATIONAL 06 Norwich Gallery, Norwich
	SACRED SELECTIONS (limited edition audio CD/publication, in association with Locus+)
2006	YOU SHALL KNOW OUR VELOCITY, Baltic Centre for Contemporary Art, Gateshead
2005	OUR SURROUNDINGS, Dundee Contemporary Arts, Dundee
	SACRED SELECTIONS (performance with Grizedale Arts as part of the Coniston Festival)
	EVERYTHING MUST GO, Workplace Gallery, Gateshead
2004	ROMANTIC DETACHMENT, PS1/MOMA, New York and Chapter Gallery, Cardiff, Wales (touring)
	QSL, FM radio broadcast, Middlesbrough Institute of Modern Art (MIMA), Middlesbrough
	S1 SALON 'CABARET', S1 Art Space, Sheffield
2003	SPACE BETWEEN US, Friar House, Newcastle-upon-Tyne (curated by SMART, Amsterdam)
	LET'S GET MARRIED TODAY, Grizedale Arts, Cumbria
	VACANT, Vardy Gallery, Sunderland
	LUX OPEN, Royal College of Art, London
2002	MEMENTO, Underground stations on the Tyne and Wear Metro system
	VideoROM, Museo Arte Contemporanea, Rome and Gallery of Modern Art, Bergamo, Italy
	LAND AND THE SAMLING, Kielder Water and Forest, Northumberland
	MOSTYN 12 OPEN, Oriel Mostyn Gallery, Llandudno, Wales
	VISIONS IN THE NUNNERY, Nunnery Gallery, Bow Arts Trust, London
2000	ROLL-IN' ALONG (solo, public video projection with Work and Leisure International, Manchester)

Matt Stokes
Stills from LONG AFTER TONIGHT (2005)
Super 16mm film / DVD, 7 mins.

Matt Stokes
Print designs for SACRED SELECTIONS (2005)
Series of live pipe organ performances
with accompanying poster and handbills
Commissioned by Dundee Contemporary Arts
Supported by The Royal College of Organists

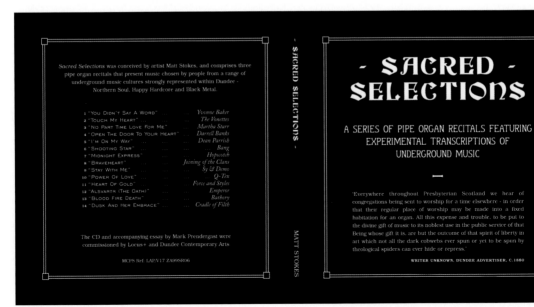

SACRED SELECTIONS

A series of organ recitals featuring experimental transcriptions of underground music

Northern Soul

WEDNESDAY 25TH MAY 2005 AT 7 O'CLOCK
ST. SALVADOR'S CHURCH, CHURCH STREET, DUNDEE

Music selected by: Ali Duff, Tam McClymont, Alan Watson
Transcribed for the organ by: Andrew Macintosh
Performed by: Andrew Macintosh (RCO) and Stuart Muir (City Organist)

Happy Hardcore

WEDNESDAY 29TH JUNE 2005 AT 7 O'CLOCK
ST. ANDREW'S CHURCH, KING STREET, DUNDEE

Music selected by DJ Sy (Quosh Records) · Transcribed for the organ by John Riley
Performed by John Riley of Forteviot.

Black Metal

WEDNESDAY 13TH JULY 2005 AT 7 O'CLOCK
CAIRD HALL, CITY SQUARE, DUNDEE

Music selected by Jess Cox · Transcribed for the organ by Andrew Macintosh
Performed by Stuart Muir / John Riley

Tickets for all events are FREE and available from Dundee Contemporary Arts

With special thanks to: Andrew Macintosh, John Riley and Stuart Muir, Ali Duff, Tam McClymont, Alan Watson, DJ Sy and Jess Cox.
Supported by The Royal College of Organists. 'Sacred Selections' is commissioned by Dundee Contemporary Arts as part of the exhibition Our Surroundings.

Dundee Contemporary Arts

152 Nethergate · Dundee DD1 4DY · Telephone 01382 909900 · www.dca.org.uk

Supported by:

SUE TOMPKINS

Sue Tompkins' solo live performances exist as strung out exercises in associative free thought: performance poetry that moves from the page to the voice, from speech to song, from song to signal, from signal to pure sound. The performances rely heavily on Tompkins' spiky energy and resilient presence. The delivery is often difficult: coos, trills, chants, and stammered semi-improvised, oddly stressed exclamations. Tompkins frequently morphs between similar-sounding phrases; words are posed as questions and repetition is key, a sentence spoken over and over until it leads to an unrelated idea.

Tompkins uses oral traditions from folk and religious vernacular to fem-punk; the subjects are immediate yet elusive, maudlin and celebratory: love, sex, everyday conversation and banal observation. Embarrassing, emotional lamentations are juxtaposed with the repetition of words such as 'Xerox' or the statement 'I never stay in hotels' or 'I read it in a magazine'. Tompkins sometimes incorporates snatches of lyrics from popular songs. No More Cola Wars (2004) for example uses a framework of Beach Boys hits. In a performance at her solo exhibition in January 2006 at The Modern Institute, Glasgow she sang (rather than spoke) — with fellow artist Alan Michael on guitar — a rendition of Richard Thompson's I Want to See The Bright Lights Tonight, Bruce Springsteen's Dancing in the Dark and David Bowie's Be My Wife — a combined selection which is classic Tompkins in terms of idiosyncratic mood, atmospheric connotation and sonic pitch.

The inventiveness and energy of recent performances More Cola Wars (2004) performed at Switchspace, Glasgow and Elephants Galore (2005) witnessed at Glasgow's CCA and at the Scottish Pavillion in Venice totally absorbed the audience. Sue always positions herself informally to the audience, fairly scruffy, foot tapping in pace with her controlled enunciation, a huge folder of paper which she uses as her only prop, alongside her vulnerability and her enigmatic grin. According to Sarah Lowndes, Glasgow-based writer and long-time chronicler of the local scene, "Tompkins' breath control, foot movements, pauses and smashed punctuation render the allusive content of the work more difficult to reach, but more ultimately far more memorable than in a standard narrative or pop song."

Tompkins' object/collage-based work may include fur, magazine cut-outs, plastic bottles, ribbons, felt-tip pen drawings or more recently text drawings made with a typewriter in which words slip and slide across the page with the staccato rhythm of her performances. The newsprint pages are folded and typed on with an old-fashioned typewriter which brings with it a particular aura and era of communication – you have to hunch over a typewriter and stab noisily at the keys. The ink can smudge. A typewriter in a lady's hands evokes L-shaped rooms and Virago classics not corporate efficiency and the 'paper-free office'. Pertinent then that you discover the word 'browse' with all its contemporary connotations deliberately folded into one wall mounted text-collage.

Often these pieces bear traces or relate directly to Tompkins' performances. Tompkins often collaborates with her twin sister Hayley and she has also worked with the collaborative Glasgow-based group Elizabeth Go, producing performance installations and short films. Tompkins was lead singer and writer in the now defunct but legendary Glasgow band Life Without Buildings. Tompkins has also collaborated with fellow artist Luke Fowler on the RD Laing inspired piece Be Dear Crazy Loud (2003).

Tompkins talks about her work in terms of 'editing', 'layering', 'arranging' and 'configuring'. In her solo show at The Modern Institute Tompkins chose to include paintings for the first time: a painting show hung over text arrangements, the paintings themselves a fresh arrangement of words and colour, statement and ground. The statements serve as mood-altering excerpts from Tompkins' performances, 'Stay Puft', 'Gleam Blue', 'Lots Plants Cheap' and 'Happy Men'. Tompkins makes patterns, repetitions and overlaps in her own linguistic register. This is an ongoing relationship of elements which she extends to the viewer through making ongoing connections between the works presented in each of the three venues for the Beck's Futures 2006. (PS)

BORN 1971 IN LEIGHTON BUZZARD. LIVES AND WORKS IN GLASGOW.

SELECTED RECENT EXHIBITIONS & PERFORMANCES

2006	WRITING IN STROBE, Dicksmith Gallery, London (solo)
	Andrew Kreps Gallery, New York (two-person show with Hayley Tompkins)
	The Modern Institute, Glasgow (solo)
2005	IN THE ZONE OF YOUR EYES, West London Projects (two person show/live performance with Hayley Tompkins)
	WHEN IT BROKE, ArtForum Berlin, Galerie Giti Nourbaksch, Berlin (two-person show with Hayley Tompkins)
	ON STAGE, Galerie Giti Nourbaksch, Berlin (live performance with Jamie Isenstein)
	SELECTIVE MEMORY opening, Scottish Pavillion, Venice Biennale, Venice (live performance ELEPHANTS GALORE)
	RADIANCE, Sorcha Dallas, Glasgow
	IN THE POEM ABOUT LOVE YOU DON'T WRITE THE WORD LOVE, CCA, Glasgow (live performance ELEPHANTS GALORE)
2004	MORE COLA WARS, Switchspace, Glasgow (live performance)
	Live performance, Museum of Garden History, London (in association with the Sonic Arts Network and Resonance FM)
	COUNTRY GRAMMAR, Doggerfisher, Edinburgh
2003	BE DEAR CRAZY LOUD, Flourish Nights, Glasgow (performance with Luke Fowler and P6)

OPPOSITE:
Sue Tompkins
Detail from UNTITLED (2004)
23 sheets of typed text on paper,
framed magazine page

BELOW (TOP):
Hayley Tompkins and Sue Tompkins
UNTITLED (2005)
Interview magazine, collage, 43.8 x 39cm

PREVIOUS PAGE (BOTTOM) AND BELOW:
Sue Tompkins
Details from installation UNTITLED (2005)
72 sheets of typewritten text on
newsprint, folded, 2 plastic Spa bottles,
1 blue, 1 red on shelf

OPPOSITE:
Hayley Tompkins and Sue Tompkins
UNTITLED (2005)
Interview magazine, collage
42.2 x 22cm

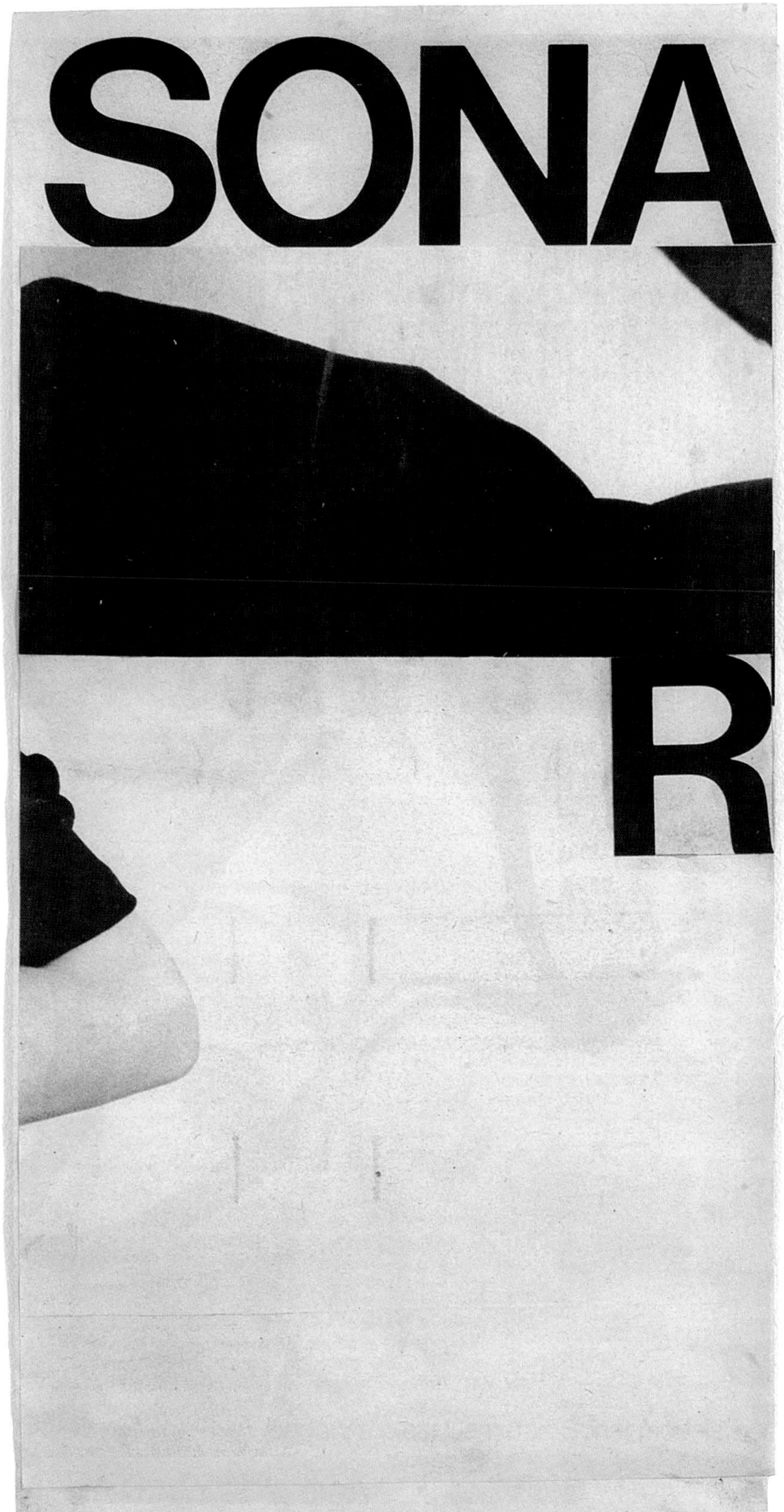
SONA
R

BEDWYR WILLIAMS

With Bedwyr Williams, as with, say, William Wegman and Sean Landers, it is futile and besides the point to try to determine where the art begins and the comedy leaves off. Besides the sculptural costumes, his performances are virtually indistinguishable from stand-up routines, while, more generally, comedy and performativity run through his entire practice, whatever the medium (installation, posters, photography, books). Williams' art/comedy is both observational and autobiographical, the humour and pathos arising from cultural misunderstandings and seemingly minor everyday humiliations. Often the source of this is his apparently contradictory identity as an itinerant contemporary artist, whose first language is Welsh, who lives in a small, distinctly uncosmopolitan town in North Wales. Provincial pathos, macho behaviour, cultural stereotyping and art world pretensions have been targets of his off-kilter satire, as have artist-in-residency schemes, unfamiliar foreign customs and the solemnity of official Welsh art.

Artists' books have been Williams' preferred means of reflecting on an extended set of circumstances. In 2002 the Arts Council of Wales invited him to document small-scale recipients of National Lottery funding. In 'Operation Ferrule', Williams consigns objectivity to the wind, his accounts of each project reading instead like diary entries or short detective stories of his encounters with various grass-roots cultural phenomena: a rag rug-making circle, a drumming workshop and a jazz society, for example. Mishap, bewilderment and affectionate satire also characterize 'Basta' (meaning 'Enough'), a bathetic photo-essay documenting his experience as artist-in-residence at the Welsh Pavilion during the most recent Venice Biennale. Its subtitle, 'When Residencies go Bad,' is one most artists at one time or another can relate to. When Williams 'critiques' institutions, he does so seemingly unselfconsciously and naïvely, in the guise of an idiot-savant.

The approach is sometimes reminiscent of Jacques Tati's Monsieur Hulot's misadventures with International Style architecture. 'Certain Welsh Artists', by contrast, is an adapted readymade: a one-off 'Nude' edition of 'Certain Welsh Artists' (by Iwan Bala), a contemporary art publication well-known in Wales, détourned, simply, with the addition of a sticker of the kind used for sale prices.

Williams has appeared in many guises in his work, which tend to function as alter-egos: as a Welsh bard with harp and flowing white beard in Bard Attitude (2005), a photographic re-make of a c.18th-century romantic painting by Thomas Jones; as the Dinghy King (2003), a fictitious marine deity in angler's waterproofs and seaside inflatables, who becomes Colwyn Bay's (a local holiday resort) nemesis; and as a politician running for office whose image appeared alongside real-life candidates on hoardings around Venice during the last Biennale.

His first major solo exhibition, at Chapter Arts Centre in Cardiff and Store in London, was an installation representing a miniature landscape made up of several interpenetrating snooker tables and model train sets, accompanied by Classic FM. In Tyranny of the Meek (2005), Williams imagines a Lilliputia in which the model train society to which he belonged as a boy belatedly exacts revenge on the intimidating members of the snooker club with whom they used to share a leisure hall. Green baizes now serve as rolling valleys and rows of sawn-up cues form bridges and aqueducts on which tiny trains pull wagons of snooker-related plunder – blue coal (bits of cue chalk) and logs (sections of cue). (AF)

BORN 1974 IN ST. ASAPH, WALES. LIVES AND WORKS IN CAERNARFON, WALES.

SELECTED RECENT EXHIBITIONS & PERFORMANCES

2005	TYRANNY OF THE MEEK, Store Gallery, London (solo)
	BROUGHT TO LIGHT, Oriel Mostyn, Llandudno, Wales
	FLOURISH, Moravian Gallery, Brno, Czech Republic
	TAKE IT FURTHER, Andrew Mummery Gallery, London
	ANIMA ANIMUS, B312 Montreal, Canada
	SOMEWHERE ELSE representing Wales at the Venice Biennale
	OVER & OVER, AGAIN & AGAIN, Contemporary Art Centre, Vilnius, Lithuania
2004	TYRANNY OF THE MEEK, Chapter, Cardiff (solo)
	ROMANTIC DETACHMENT, PS1/MOMA New York, and Grizedale Arts (touring)
	NOW THEN, NOW THEN, The International3, Manchester
	SUMMER INVITATIONAL, fa projects, London
2003	APROPOS OF NOTHING, G39, Cardiff
	LET'S GET MARRIED TODAY, Grizedale Arts, Cumbria
	ANGLO PONCE, The Trade Apartment, London
	BLAENAU VISTA SOCIAL CLUB (part of ROADSHOW, a Grizedale Arts project)
	TALWRN CELF (ART COCKFIGHT), National Eisteddfod, Wales (performance)
2002	OPERATION FERRULE, Fotogallery Cardiff (touring until 2003, solo)
	MISTAR AR MISTAR MOSTYN, Oriel Mostyn, Llandudno, Wales (solo)

ABOVE: Bedwyr Williams
ALBATROSS (2005)
Still from live performance,
National Eisteddfod, Faenol

OPPOSITE: Bedwyr Williams
BLAENAU VISTA SOCIAL CLUB (2003)
Installation at Grizedale Roadshow,
Blaenau Ffestiniog, Wales

OPPOSITE: Bedwyr Williams
BARD ATTITUDE (2005)
Photographic print

Bedwyr Williams
TYRANNY OF THE MEEK (2004)
Installation views, Chapter Visual Arts

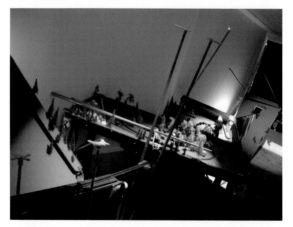

STANDING STILL
AND MOVING FORWARD
BY ROB BOWMAN

Nobody's perfect. As a statement, it's a seemingly less than perfect starting point for a text related to a major award for UK-based visual artists. Amid the hyperbole that often surrounds prizes and awards, it feels like the rhetorical equivalent of taking aim, firing, missing, and then merely shrugging.

In the field of the visual arts, and particularly for institutions in any way committed to 'the up-and-coming', to supporting the more innovative activities within contemporary culture, one of the things consistently at stake is the question of how to frame work that is still developing, and coherently 'showcase' that which may not yet be neatly packaged, categorised, historicised; neither perfectly understood, nor understood as perfect. It is arguably one of the functions of such institutions to offer just such a framework for evolving and more dynamic areas of art practice. It would likely be a structure within which artists' work could wave its arms and be recognised as doing exactly that: waving, not drowning. The format might allow for the illustration of connections between different artists' approaches to making work in a continually changing context. And, it would enable the presentation of that work to an audience: a public whose visual literacy is undeniably on the increase, and press and critics whose reactions to contemporary art sometimes seem — to artists, if to no-one else — like the mood swings of an hormonal teenager.

There is a problem, however. Even when embracing the more experimental, or actively seeking out a group of artists with 'future potential' there is a risk of being seen to claim definitude. A process of selection for an award that might have been about identifying the most interesting and outstanding practitioners amongst currently emerging artists is viewed as a prediction of the future of art itself. For better or for worse, it is a position that has been attributed to Beck's Futures in the past. On countback, the phrase 'I've seen the Future and. ', comfortably makes the Top Ten of subeditors' headline-reviews of the exhibition. I am not stating the case for its re-use. On the contrary, what has been offered has been under the banner of the most interesting and innovative of current, emergent practice: those whose work suggests to selectors that they, above others put forward, are likely to continue to develop their work in exciting and relevant ways in the future. That continues to be the case.

The view of selectors — and the panel changes from year to year — does of course offer this validation to artists' work, but there has been an additional, new element in this year's selection process. A panel of researchers, consisting of eight arts-professionals, was asked to pull together a long-list of nominations. The process is from the outset independent and, unlike in previous years, makes visible a form of benchmarking by arts professionals throughout the process. There is an undoubted level of recognition for artists put forward, even if it is not visible, even before short-listing begins.

Beck's Futures is — or rather was — a single, annual exhibition (even if it previously travelled in the UK) as well as being the singular outcome of a selection procedure that enjoys credibility because of the standing and experience of those invited. But being involved in the selection has also inevitably confronted judges with questions of relative value, and of how to arrive at the definitive judgements they are called upon to make. If the process is seen more as a way of opening doors for those most promising UK-based artists than it is a way of pinning down an exact and precise future through the act of selection, then the exhibition subsequently realised is even more obviously only one of a number of possible results. It seems sympathetic to this predicament that the shortlist for Beck's Futures 2006 does signify one act of selection – one shortlist of thirteen artists – but at the same time acknowledges that the exhibition could take a number of forms: at least two other forms in two further venues besides the ICA. It reflects what has not perhaps previously been articulated. Where people might have felt confident about the exhibition's supposed intention to give a singular and definitive view of art's future, the possibilities have always been grounded within artists' present activity. Even within the here and now, the possibilities are still many and various.

To add further fuel to the question of what definitive values it might be possible to assign to artists' individual and collective contributions to the exhibition, a public vote operates this year, designed to contribute to the identification of an overall award-winner. The public 'block' vote is an additional, seventh voice alongside those on the judging panel. An invitation is given to visitors – who will in all likelihood each see only one of the three exhibitions being staged across the UK – to express feelings, and identify favourites. Three different possible experiences, seen through countless different pairs of eyes, is not a strong argument for the validity of a single definitive judgement. On the contrary, it would seem to reveal something more fundamental about the process by which it is possible to make such evaluations.

All the while, both public and media seem consistently to ask that those visible in creative fields – not only artists but writers, singers, filmmakers, and assorted other 'names' doing assorted things – should demonstrate excellence, and aspire to be the best in their chosen field. We enjoy ever more visible, and frequent news of the achievements of those who pick up a Mobo, Oscar, or the Celebrity Chef of the Year Award. Nobody likes a failure or a reject; and there are many that are quickly rejected, even if failure also seems to turn out to be a reasonable basis for a future 'career'. Is it all just a natural projection of our own competitive aspirations to outdo each other?

Whether that's the case or not, guessing at the criteria for the host of awards for those within serious creative disciplines isn't difficult. We can safely assume that the judges/selectors/academy are looking for what they consider to be stand-out excellence, originality and brilliance within the field, either demonstrated in the past or promised for the future. In the visual arts, the work that forms the basis on which awards are allocated certainly doesn't have to be rational even if it has a rationale. It can be wildly, unrestrainedly emotional, emotive, or funny – as long as the work is more effective than possible alternatives in getting at something pertinent. It's the original and thought provoking that gains support; in the crassest possible terms, a case of 'the best of the best, and forget the rest'. But 'the rest' is also part of the story, and even more so for this year's Beck's Futures. If there appears to be a healthy level of productivity across the visual arts then that is surely something to be celebrated, and perhaps even examined more closely. It is stating the obvious to note that the participation of those nominated to the awards provides Beck's Futures with its life-blood, despite the risk to participants that they might (erroneously, because of the recognition signalled by their very nomination) be seen as 'also-rans'. What could be more revealing of the current climate for, and the shifts in, contemporary art practice than a glimpse of the wider field of activity? If one of the functions of public arts institutions is to offer insight into what artists are currently doing through activities that aim to encourage engagement with the work of the most interesting and innovative artists, then this should not discourage us from scraping at the surface and trying to find out the wider context for that activity. Beck's Futures' 'back-story' – the process of nomination – provides the opportunity for exactly that process of discovery. This is partly the logic, even if for personal reasons some chose not to be represented, behind including in this catalogue brief statements about the work of the artists proposed but not short-listed this year.

One of a long list of possible titles for this year's Beck's Futures exhibitions, offered to the curators by the short-listed artist Stefan Brüggemann as part of his contribution to the exhibition, was 'How did we get here?' It does not sound like a question asked with much expectation of being answered. An insight into the back-story, though, alongside a consideration of the works by those participating in this year's exhibition, can only increase our chances of establishing where we have actually got to, even if we're not yet sure how we got here. It might even suggest we should look forward to tomorrow.

ROB BOWMAN
CURATOR, ICA EXHIBITIONS

BECK'S FUTURES 2006
NOMINATED ARTISTS

KARLA BLACK
SIMON BLACKMORE
TOM AND SIMON BLOOR
MATT BRYANS
TOMAS CHAFFE
RUTH CLAXTON
LUKE COLLINS
WILLIAM DANIELS
KATE DAVIS
WILL DUKE
MARTE EKNAES
ERICA EYRES
LAURENCE FIGGIS
JOSEPHINE FLYNN
RACHEL GOODYEAR
KATE OWENS AND TOMMY GRACE
ANTHEA HAMILTON
LUCY HARRISON
KATHLEEN HERBERT
DEAN HUGHES

I-CABIN
TORSTEN LAUSCHMANN
PETER LIVERSIDGE
CAMILLA LØW
TIM MACHIN
HEATHER AND IVAN MORISON
JOHN MULLEN AND LEE O'CONNOR
KATIE ORTON
MICK PETER
ALEX POLLARD
MARY REDMOND
HANNAH RICKARDS
ANTHONY SHAPLAND

KARLA BLACK

Karla Black is an articulate and informed artist, Robert Smithson, Robert Morris and Robert Rauschenberg all having played a role in the development of her work. Yet her current practice pulls free from the historical wake of these seminal '60s and '70s male figures in order to create a distinctive body of process-based sculpture more attuned to the agenda of Carolee Schneemann. Constructed from materials that are often domestic and 'uncultured', either intrinsically (Vaseline, clothing, and flour) or in terms of treatment (powdered plaster thrown on the floor), her works are as much concerned with the process of their own making as they are with their final form. Her sculptures retain a visible record of their own production, and Black regards each of them as residual performance as much as autonomous article. Her works might almost be described as 'coming-into-being'; in daring to be very little – almost nothing – Karla Black's open objects achieve maximum psychic resonance and improvisational freedom. (JC)

Born 1972 in Alexandria, Scotland. Lives and works in Glasgow.
SELECTED RECENT EXHIBITIONS
2006   KARLA BLACK, Broadway 1602, New York
        KARLA BLACK, Mary Mary, Glasgow
2004   CHEYNEY THOMPSON/KARLA BLACK, Transmission Gallery, Glasgow
2003   EAST INTERNATIONAL, Norwich Gallery, Norwich

SIMON BLACKMORE

Simon Blackmore brings a playful and enquiring mind to the world of creative technology. His recent performative experiments downplay the often complex production processes he sets up in order to produce low-tech responses to the environment. Weather Guitar, for example, is a robotic musical instrument that responds to changes in the weather; a moody presence, it interprets shifts in atmosphere to create spontaneous, electronically 'ad-libbed' compositions. Another project, Sprite Musketeer, takes the form of a 1970s caravan converted into a mobile cinema space and framing device. Parked in areas of Wales or the Lake District depicted by Romantic artists, it provides an alternative way of experiencing those now famously Sublime landscapes. While both projects indicate an engaging 'geekiness' – a compulsion to mess about with objects, electronics and sounds – Simon Blackmore's work is surprisingly emotive, posing profound questions about the role of the individual in a technically driven society. (CD)

Born 1976 in Honiton, Devon. Lives and works in Manchester.
SELECTED RECENT EXHIBITIONS
2005   WEATHER GUITAR, Ikon, Birmingham (solo)
        ULTRASOUND, Huddersfield (performance with Owl Project)
        WILL HAVE DONE, Mill Workers, Manchester
        GATHERING MOSS, Q Arts, Derby (performance with Owl Project)

TOM AND SIMON BLOOR

The Bloor brothers' main focus is communicating information. With their contemporary design sensibility, they pull off a strange mix of retro materials and pop production values, combining these with cheap, accessible distribution circuits in order to disseminate hidden connections and obscure histories. Making use of a strong graphic coherence, they forge temporary links between such diverse elements as the Lunar Society (an 18th-century Birmingham-based philosophical group) and the rock band Black Sabbath. Their practice takes in stickers, posters, flyers, wallpapers, neon signs, pamphlets and books. The project What Are the Senses? (2005) examined the work of Modernist architect Berthold Lubetkin at Dudley Zoo during the 1930s. As well as conventional documentation, the artists collected historical material such as advertisements that feature the zoo. During their installation there, they took over two listed kiosks, which they fly-posted with relevant texts and images, and from which they distributed a booklet illustrating their research. They now plan to use the kiosks' display boards to make what they call 'nomadic furniture'. (FM)

Both born 1973 in Birmingham. Live and work in Birmingham.
SELECTED RECENT EXHIBITIONS
2005   PUBLIC STRUCTURES, 2nd GuangZhou Triennial, China
              ART SHEFFIELD 05, SPECTATOR T, S1 Artspace & other venues, Sheffield
              WHAT ARE THE SENSES?, Dudley Zoological Gardens, Dudley (solo)
2004   ROMANTIC DETACHMENT, PS1/MOMA, New York, and Grizedale Arts (touring)

MATT BRYANS

Stuck directly to gallery walls, Matt Bryans' collages look from a distance like charcoal drawings, with eyes and mouths emerging from the faces of ghouls; closer inspection, however, reveals the works to be made from fragments of newspaper photographs. Abstracted in this way, his 'faces' have a melancholic Symbolist air, yet their features clearly belong to specific people: before Bryans got to them, many were well known to millions of readers. Others would be identifiable by their accompanying stories, which the readers might recall. In this way, Bryans' collages could be perceived as a kind of journalistic afterlife — images from the news that re-surface in dreams or distant memories. In contrast to their mechanically reproduced sources, his works feel almost organic, like spreading creepers. Bryans also produces single images in the same way, and these belong to one of three traditional genres: portraits with a strong 19th-century flavour, Turneresque landscapes, and architecture that is faintly reminiscent of film sets. (AF)

Born 1977 in Croydon. Lives and works in London.
SELECTED RECENT EXHIBITIONS
2006   MATT BRYANS, Kate MacGarry, London (solo)
2005   PICTURE THIS!, MMD (Museum Dhondt-Dhaenens) Deurle, Belgium (solo)
              RECENT ACQUISITIONS, Museum of Contemporary Art, Los Angeles
              PIN-UP, Tate Modern (with Dr Lakra, Sebastiaan Bremer, Amie Dicke,
              Godfried Donkor, Wangechi Mutu, Jocum Nordstrom & Steven Shearer & Nicole Wermers)

TOMAS CHAFFE

Tomas Chaffe carefully observes the co-ordinates, parameters and operative structures that define and regulate museums and art galleries, working within a wide range of media including drawing, video and sculpture. Chaffe uses these explorations not so much to formulate a clear opposition to the settings in which art is commonly shown, but to pose questions and suggest subtle shifts in the hope of provoking change. A central element of Chaffe's work involves an ongoing series of drawings depicting museums and gallery spaces in states of disorder, through which he challenges the conditions in which art is conventionally shown. The drawings function in two ways: both as working proposals and, more importantly, as exhibitable elements in their own individual right. Although some works express ideas that he hopes will one day be realised, others outline concepts that, for political, economic or practical reasons, can never be brought to fruition; in this way, he uses art to transpose notions of impossibility and failure into interesting and productive prospects. (AC)

Born 1980 in Nottingham. Lives and works in Nottingham.

SELECTED RECENT EXHIBITIONS

2005   POST NOTES, Midway Contemporary Art, Minneapolis and ICA, London

2004   POST NOTES SHOW, Floating IP, Manchester

2003   EVE OF DESTRUCTION, The International3, Manchester
        Fresh Art, London

RUTH CLAXTON

Ruth Claxton's I thought I was the Audience and then I looked at You is an on-going installation featuring ceramic figurines adorned with garish ribbons and baubles. The artist places these 'adjusted' objects on a variety of surfaces (such as tables and chairs) to create a landscape populated by figures, as she puts it, 'blindly cosseted in isolating balls of suffocating prettiness' Claxton's installations are resolutely sculptural, but the relationship between their constituent parts takes them beyond an indulgence in found kitsch; her figures look out over a vista from which they are shielded, so they are transformed from decorative objects into players in an interactive environment. While the artist's work has clear links with the '70s feminist re-appropriations of domestic objects, they also reflect her fresh perspective on a consuming gaze that, although visually exuberant, is precisely plotted. (CD)

Born 1971 in Ipswich. Lives and works in Birmingham.

SELECTED RECENT EXHIBITIONS

2006   RUTH CLAXTON AND AMY MARLETTA, The Collective, Edinburgh
        DRAWING LINKS, The Drawing Room, London (selected by Helen Legg)

2004   I THOUGHT I WAS THE AUDIENCE AND THEN I LOOKED AT YOU, University Gallery, Colchester (solo)

2003   EAST INTERNATIONAL, Norwich Gallery, Norwich (selected by Toby Webster and Eva Rothschild)

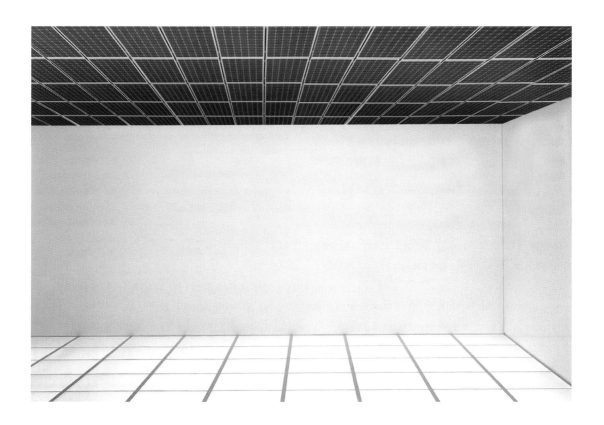

LUKE COLLINS

For Luke Collins' BA degree show, he installed a striking series of windsocks, a murky video of Willow the Wisp and a black-and-white tiled armature — all part of a carefully constructed universe that reflects his performative fascination with myth-building. Exploring this theme further, he collaborated with Stephen Murray on Seek Ye Ur Questing Beast (2005) at Intermedia, Glasgow. Here, a prescribed route through the installation began with a rough model of the Babylonian Ishtar Gate. Built c.575BC, the original was used as an army camp during the invasion of Iraq, while the Intermedia version represents the fall of Babylon in a downsized form for the TV eye. Further along the route was Crest, a talisman made from a woman's glittery glove, a feather, a ribbon, a bell and a duster. Through the magic of cheap special effects, the glove became a mouth and repeated the incantation 'enchant me baby' — the essence of ur questing beast, a role-playing masculinity as improvised psychodrama. (NM)

Born 1979 in London. Lives and works in Glasgow.
SELECTED RECENT EXHIBITIONS
2006   CAPTURED, Tramway, Glasgow
        CHATEAU MATEAU, The Old Carpet Factory, Glasgow
2004   THE GARDEN, Castlefield Gallery, Manchester
2003   SEEK YE UR QUESTING BEAST, Intermedia, Glasgow (with Stephen Murray)

WILLIAM DANIELS

William Daniels produces photorealistic paintings of his own torn-card models, which in turn are inspired by the work of masters as diverse as Caravaggio, Cézanne and Baselitz. The models are all quite similar, and look as if they were assembled quickly — much more quickly than the paintings, whose hyper-realism is so astonishing that we want to touch it to make sure it is painted — that the artist hasn't stuck on bits of card. The rough construction of Daniels' models and the modest dimensions of his paintings both suggest that the original works are less profound and accessible than they once were — as if we live in an age without belief: in religion, ideology or art. Although his lumpen forms and scrupulous brushwork both serve, in contradictory ways, to denigrate these originals, they also highlight a shared creative core: intense observation. Daniels makes the play of light on crumpled card as vivid and magical as the flesh of Christ, a turbulent sea, or a mountain in Provençe. (AF)

Born 1976 in Brighton. Lives and works in London.
SELECTED RECENT EXHIBITIONS
2006   WILLIAMS DANIELS & FIONA JARDINE, Transmission, Glasgow
2005   Vilma Gold, London (solo)
        Marc Foxx, Los Angeles (solo)
        WASTE MATERIAL, The Drawing Room, London

KATE DAVIS

It's not simply the depicted relationships between people and objects that count in Kate Davis' work: it's also the relations viewers have with them in the shared space of the gallery. To view Davis' images in situ, visitors are often hampered by obstacles: a bicycle or raised floor, for example. As a result, the act of looking suddenly becomes physically uncomfortable, socially awkward and mentally self-conscious. The images themselves frequently explore these unsettling interactions, with uncanny metamorphoses of people into objects (and vice versa).

Echoes of Ernst, Brancusi, Picasso, and Léger reverberate in Davis' work, yet the conventional masculine values these Masters often exemplify – power, certitude, authority, legitimacy – do not always prevail in her world. In order to experience this work fully, viewers need to readjust habitual modes of thought and behaviour. In a way, the testing of rules and conventions to their breaking point is at the heart of Davis' practice, yet she produces work of disarming subtlety, deceptive lightness and seductive simplicity. (JC)

Born 1977 in Wellington, New Zealand. Lives and works in Glasgow.
SELECTED RECENT EXHIBITIONS
2006   Dicksmith Gallery, London (solo)
        Kunsthalle Basel, Switzerland (solo)
2005   The Breeder Project, Athens (solo)
        MUSIC OF THE FUTURE, Gasworks, London (curated by Laurence Kavanagh)
2004   COUNTRY GRAMMAR, The Gallery of Modern Art, Glasgow

WILL DUKE

Although Will Duke's primary field – digital animation – is still in the early stages of development as an artistic form, he has already found a well-articulated and distinctive 'voice'. His immediate and hypnotic pieces suspend viewers somewhere between spellbound movie- watching and tense video-game playing. Duke's virtual spaces are deserted, yet loaded with ominous foreboding, an effect created by careful control of time, duration and pace. Movement through these spaces is sometimes a seamless and hallucinatory horizontal drift, at other times a violent and

vertiginous swoop. Dramatic use of sound and light also contributes to the exceptionally powerful impact of this work when installed. Projected on a large scale in the relatively small, dark space at Market Gallery in Glasgow, We fashioned the city on stolen memories (2005) was a scary proposition. Far from being a nerdy indulgence in new technology, this, like other examples of Duke's work, uses that technology to comment on a world it renders increasingly abstract, unreal and inhuman. (JC)

Born 1973 in Chichester. Lives and works in Glasgow.
SELECTED RECENT EXHIBITIONS
2006   NEW WORK SCOTLAND PROGRAMME, Collective Gallery, Edinburgh
2005   NADA Art Fair (with Transmission Gallery), Miami, USA
        SCHIMATIZO 2005, London
        REVISIONS IN RELATIVE TIME, Market Gallery, Glasgow

MARTE EKNAES

Marked by its concern for surface, light, disappearance, and the limits of materials, the work of Norwegian-born Marte Eknaes stakes out the potential for a practice rooted in representation to be critical. Profoundly influenced by her extended residence in southern California, Eknaes mines the alternately sadistic and symbiotic relationship between image and architectural surface that defines the contemporary built environment. The manageable scale of her contaminated Minimalism suggests a series of improvisational maquettes, jagged proposals whose use-value remains uncertain. Using degraded materials, she constructs metallic dystopias in miniature; her materials approximate machine-age metal, but such industrial swagger is thwarted by the reality that the pieces are often made from Perspex, tape, and mirror card. There are often tensions between various elements in her work, and these tensions delineate spaces of unease and frustration — sensibilities very much in tune with the increasing uncertainty of both the institutional and urban contexts that frame her practice. (SC)

Born 1978 in Elverum, Norway. Lives and works in London.
SELECTED RECENT EXHIBITIONS
2005   MODIFIED UNIFORMS, Blum & Poe, Los Angeles
          STRAGEDY, Art2102, Los Angeles
2004   RESTLESS LOOKS, Gallery Schaufenster, Oslo
2003   EMERGENCE, Track 16 Gallery, Santa Monica

ERICA EYRES

Often working in the potentially banal medium of confessional video, Canadian-born Erica Eyres nonetheless produces films that are genuinely unsettling, utterly compelling and unavoidably memorable. Technically and formally, her videos frequently borrow from the simple and direct effects of low budget TV, but with an added psychological complexity so acute that viewers don't know whether to laugh, recoil, or call the emergency services. Employing unvarnished artificiality, alarming candour and — especially in her drawings — a sense of awkward sexual vulnerability, Eyres leads us to a kind of truth. Like the pitiful and hilarious women in her work, we are all struggling with various handed-down ways of discovering identity and self-worth. What is happening 'inside' us can seem mismatched with external expectations and demands. Where are love and fulfilment hiding in this world of false appearances and contradictory appeals? How much physical and emotional damage must we suffer to get by? (JC)

Born 1980 in Winnipeg, Canada. Lives and works in Glasgow.
SELECTED RECENT EXHIBITIONS
2006   Rokeby, London (solo, forthcoming)
2005   BLOOMBERG NEW CONTEMPORARIES, Barbican, London
2004   'NEATH THE GREEN GREEN GRASS, Mary Mary, Glasgow (solo)
          IT'S FOR THE BEST, Dwarf Gallery, Reykjavik (solo)
          CARDBOARD FACTORY, Leisure Club Mogadishni, Copenhagen

LAURENCE FIGGIS

Science suggests the possibility of parallel worlds — worlds governed by other laws, other logics. Looking at Laurence Figgis' work is a little like entering such a parallel universe. His seminal piece The Great Macguffin (2005), shown at Transmission in Glasgow, was made up of large, frieze-like drawings composed of seemingly sequential frames, like a vast comic book spread across the wall. Meticulously drawn, they instantly immersed the viewer in their epic structure. Eschewing a single narrative thread, they offered simultaneous and multiple choices, spreading vertically, horizontally, laterally, diagonally, forwards and backwards. If viewers lost track of one particular protagonist or object, they could follow the trajectory of some other image. If they never quite knew where they were, they never felt totally lost either. Among the number of today's young artists who construct complex narrative and symbolic systems to suggest alternative worlds and deeper realities, Laurence Figgis is unquestionably one of the most convincing and compelling. (JC)

Born 1977 in London. Lives and works in Glasgow.
SELECTED RECENT EXHIBITIONS
2006   FLYING/STEALING, Galleria S.A.L.E.S., Rome (solo)
2005   Mary Mary, Glasgow (with William Horner)
          THE GREAT MACGUFFIN, Transmission Gallery, Glasgow (solo)
2004   RED BUS GO, 273, Glasgow (with Nick Evans)

JOSEPHINE FLYNN

In New Contemporaries 2005, Josephine Flynn's cap-and-gown graduation photograph was easily the most interesting object. She had wrapped her hands in bandages for the ceremony, sabotaging the customary handshake-for-certificate ritual. On noticing this, the VIP involved whispered in her ear, "I'll just touch it, if you don't mind!" The image records a grinning Flynn proudly grasping her certificate between bandaged mittens. This portrait is neither a found photograph that has been altered, nor a re-enactment, nor a document of a performance. Flynn's blatant self-exposure is not only witty, it also has something profound to say about the staging of rites of passage, and the camera's role in documenting a life history. Still at an early stage in her career, Flynn shows great potential that is particularly evident in her recent video works. Scuff 3 is a tightly edited mix of pop, animation and tabloid imagery; flitting from fast food to global politics, it typifies the artist's humorous take on received social norms. (CD)

Born 1975 in Leigh, Greater Manchester. Lives and works in Leeds.
SELECTED RECENT EXHIBITIONS
2005   ART SHEFFIELD 05, SPECTATOR T, various venues, Sheffield
          Poster Project Commission (Cornerhouse 20th Anniversary), Cornerhouse, Manchester
          NEW CONTEMPORARIES 2005, Cornerhouse, Manchester; Spike Island & LOT, Bristol;
          Curve Gallery, Barbican, London
2004   NOW THEN, NOW THEN, The International3, Manchester

RACHEL GOODYEAR

Working principally in the medium of drawing, Rachel Goodyear depicts peculiar situations; informed by the everyday, they are also absurd and horrifying in an almost filmic way — like a frog in a jam jar by a reading lamp, or a squirrel's tail hanging out of a man's mouth. For some time, she has been constructing a fictional narrative in which the same characters re-appear and enter new scenarios, all of which involve pain, shock and wonder. In each work, the elements are articulated and juxtaposed so they can be interpreted in different ways. Goodyear pays close attention to her subjects, drawing them in a quietly beautiful and whimsical manner; sometimes, ephemeral and found artefacts such as bus tickets, paper bags, and receipts are added. When the drawings are exhibited, few of them are framed or mounted. Instead, they are positioned closely together on the gallery wall in the order they were produced: individually valid, but also part of a story. (AC)

Born 1978 in Oldham. Lives and works in Manchester.
SELECTED RECENT EXHIBITIONS
2005   ARTFUTURES, Bloomberg Space, London
        WE WILL CANCEL YOUR MEMBERSHIP AND YOU WILL OWE US NOTHING,
        The International3, Manchester (solo)
        MOSTYN OPEN, Oriel Mostyn Gallery, Llandudno
        POST NOTES, Midway Contemporary Art, Minneapolis and ICA, London

KATE OWENS AND TOMMY GRACE

Over the past five years, the individual interests and approaches of Edinburgh artists Kate Owens and Tommy Grace have been dovetailing around aspirational interiors and dusty architectural follies. In contrast to the International Modernist architecture and design with which Glasgow artists are associated, the work of Owens and Grace takes its cues from the mongrel style of the east coast of Scotland (Lothians and the Kingdom of Fife): an architectural Baroque they filter through everyday experiences and materials (Lambrini in the park), outdated optical-tricks such as 3D, romantic ghost-stories and tacky interiors source material (like Peter York's latest book Dictator's Homes). Recent installations include a collaborative contribution to the group show The Garden, at Castlefield Gallery, Manchester. Owens and Grace are founding committee members and curators at Edinburgh independent galleries Win Together and The Embassy, for whom they organise major video and performance events. (NM)

Born 1979 in Falkirk / 1979 in Edinburgh. Both live and work in Edinburgh.
SELECTED RECENT EXHIBITIONS
2006   I-RESIDENCE, Royal Scottish Academy, Edinburgh
2005   FUNCTION FORM FOLLOWS, Cooper Gallery, Dundee
        THE GARDEN, Castlefield Gallery, Manchester
        HERO OF THE TWO WORLDS, Collective Gallery Project Room, Edinburgh (Owens)
        THE ROSE TATTOO, Collective Gallery Project Room, Edinburgh (Grace)
        GATES OF ADES, Collective Gallery off-site project, Edinburgh (Owens)

ANTHEA HAMILTON

A recent RCA graduate, Anthea Hamilton creates installations from autonomous elements that she recombines to suit each exhibition space, thus forming impermanent three-dimensional collages with a slightly Modernist resonance. Hamilton's work is underpinned by a knowledgeable reworking of art history and pop culture, combined with an exploration of love and sexual desire. Materials play a crucial role in Hamilton's practice; they are part of a seductive process that draws the viewer into an imaginary flirtation. Sculptures, images, plants, drawings, and clothes all form part of the artist's vocabulary, while copper, crystals, beeswax and clay are selected for their physical appeal. In her work, Hamilton traces a personal history of love and attraction: for her, William Blake, Robbie Williams, Captain Ahab, Auguste Rodin, Woody Allen, Gustave Courbet and the Bee Gees all inhabit the distance between individuals and the object of their desire. (FM)

Born 1978 in London. Lives and works in London.
SELECTED RECENT EXHIBITIONS
2006   VIDEO LONDON, Espai Ubu, Barcelona (curated by Keith Patrick and Jennifer Thatcher)
        ICH BIN EINE SACHENAÜFSHIESSE, i-cabin, London
        SV05, Studio Voltaire, London (selected by Enrico David and Catherine Wood)
2002   FROM TARZAN TO RAMBO: IN FOCUS, Tate Modern, London

LUCY HARRISON

Lucy Harrison's recent practice examines the subjective nature of the urban experience, and involves collecting found and often defunct descriptions of cities. Centred on forgotten or ignored texts, she deals with the gaps between coded presentations and texts, and with the edges of experience around a subject of focus. Shadows; Jonathan Miles, Lecture Theatre One, Royal College of Art, November 1997 (1999) is a slide projection consisting of all the words or phrases spoken during a lecture that were not part of the script. Good Quote; Art School Marginalia (2003) is a collection of book-margin notes made in a college library. More recent projects have included include Guided Tour (Riga), a photographic and textual documentation of three city-tour walks recommended in a Soviet-era guidebook, and Fantastic Cities, in which 25 artists write about a city they have never visited, using information gleaned from fiction, film or anecdotes. Harrison's recent published works are intended for use by travellers; they are tools for negotiating urban space. (CD)

Born 1974 in Crawley, West Sussex. Lives and works in London.
SELECTED RECENT EXHIBITIONS
2006   TRES RICHES HEURES, Lokaal 01, Breda, Le BLAC, Brussels & Keith Talent Gallery, London
        LA CHEBA DEI MATI, San Servolo, Venice (cat.)
2005   AFTER THE FACT, Tullie House Museum and Art Gallery, Carlisle (cat.)
2004   FAMILY SAGAS, Northern Gallery for Contemporary Art, Sunderland

KATHLEEN HERBERT

The work of Kathleen Herbert draws on the conventions of documentary film-making to build up narratives enhanced by her sensitive and observational style. Often by redefining location and scale or layering detail, she creates intrigue, never exposing the full extent of her story. In 2003, she investigated Bletchley Park, the World War II coding-breaking centre, to produce a video work, Station X. During her interviews with surviving code-breakers, she filmed only their mouths and necks, which became increasingly strained as the impact of secrecy on their lives was revealed. Commissioned to produce a new work for Bristol in 2005, Herbert took a three-day voyage on a cargo ship from Antwerp to Bristol with a male crew of 28. The resulting 8-monitor installation highlights the solitude, monotony and disorientation experienced by contemporary seafarers. Just as Herbert's practice is marked by a cautious and sensitive research process, her work is quiet and slow burning, with a lasting and profound impact. (CD)

Born 1973 in Watford. Lives and works in London.
SELECTED RECENT EXHIBITIONS
2005   THINKING OF THE OUTSIDE: NEW ART AND THE CITY OF BRISTOL (with Nathan
        Coley, Phil Collins, Susan Hiller and Silke Otto-Knapp), multiple sites, Bristol
2004   AUCKLAND TRIENNIAL, Auckland Art Gallery, New Zealand
2002   HEIMLICH/UNHEIMLICH, RMIT Gallery, Melbourne (curated by Juliana Engberg)
        SCAPE, Art & Industry Biennial 2002, Christchurch, New Zealand

DEAN HUGHES

Dean Hughes uses ordinary items of stationery – paper, staples, gumstrip, string – to produce works that are at once drawings, sculptures and Readymades. Possessing all the formality of Minimalism and Conceptualism, they also have a whimsical accessibility that narrows the gap between life and art, and helps us to look at things in fresh new ways. Hughes' work suggests the perceptions of someone newly arrived from another planet. Seemingly unfamiliar with the objects we use to organise ourselves and others, he assigns his own poetic functions to them: in Staples pushed through paper (2002), for example, hundreds of unused staples create a rhythmic pattern on a sheet of A4, while their straight prongs cause the work to 'float' above the gallery floor. The simplicity of the artist's practice conceals the considerable labour that goes into it, and by implication, subverts the logic of our overly administered world. (AF)

Born 1974 in Manchester. Lives and works in Manchester.
SELECTED RECENT EXHIBITIONS
2004   LIVING DUST, Norwich Gallery, Norwich (curated by David Musgrave)
2003   Jack Hanley Gallery, San Francisco (solo)
2002   38 Langham St, London (solo)
        GESTURES IN SITU, Arcadia University Gallery, Pennsylvania
2001   NOTHING, Northern Gallery for Contemporary Art, Sunderland (touring)

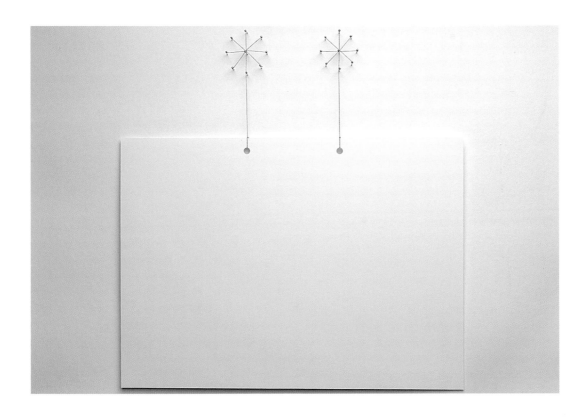

I-CABIN

Conceived as a single artistic gesture, i-cabin is both a conceptual project space and an unconventional centre of art production. Unlike traditional artist-run spaces, i-cabin is a medium for the collective artistic interventions of the three artists who run it. These 'interferences' begin with a set of rules they impose on themselves and exhibiting artists — rules that combines playfulness with rigour in the way they experiment with gallery codes and conventions. The programme itself is divided into two strands, running concurrently:

Year One involves the presentation of young artists, who are asked to deal with the gallery's rules, respond to them, and perhaps even push them further. Year Two, conceived for more established artists, encourages projects that expand beyond familiar ways of working. i-cabin infiltrates the system of art production and consumption like a virus; working with it in a complicit and critical way, the project elaborates extensively on the concept of 'artist as gallerist'. (FM)

i-cabin was established in 2004 by Juliette Blightman, Sebastian Craig and Will Morrow. Based in London (www.i-cabin.co.uk).
SELECTED RECENT EXHIBITIONS
2006   ADRIAN HERMANIDES: BLACKSON, i-cabin project space, London
2005   TILL EXIT: STRAND (IN SEVERAL HARMONIES), i-cabin project space, London
        ALEC STEADMAN: CLARENDON BUILDING PROJECT, i-cabin project space, London
        B.Y.O/RESOURCE, i-cabin project space, London

TORSTEN LAUSCHMANN

Torsten Lauschmann produces a vastly diverse range of work reprocessed for a bewildering variety of venues and audiences. In addition to his extraordinary one-person show at Transmission in 2004, he has contributed to numerous group shows, undertaken research on scratch audio video, performed as a VJ and solar-powered busker, and designed innovative software and websites, which he makes freely available. His playful approach to the use and misuse of digital technology has been highly

influential, and his great generosity has galvanised the Glasgow art world for many years. Lauschmann's recent World Jump Day scam (as Professor Hans Peter Niesward from the Institute for Gravitational Physics in Munich) was the most talked about viral jam of 2005, encouraging scientists and bloggers from around the world to discuss the physics of WJD's plan to shift the Earth's orbit and reverse global warming. (NM)

Born 1970 in Bad Soden, Germany. Lives and works in Glasgow.
SELECTED RECENT EXHIBITIONS
2005   RADIODAYS, De Appel, Amsterdam
        THE CHATEAUX at Halloween Society, ICA, London
2004   Transmission Gallery, Glasgow (solo)
2003   ZENOMAP, Venice Biennale, 2003

PETER LIVERSIDGE

Peter Liversidge's work centres on two ongoing series: one devoted to the plains of Montana, and the other to luxury consumer goods. Apparently antithetical, these concepts both inhabit the world of fantasy: Liversidge has never been to Montana, nor is he wealthy. This artist is essentially Conceptual; his lack of technical facility reduces his subject matter to the level of his whims. In Watches of Switzerland (2003), for instance, the wobbly lines in 169 paintings of Swiss watches undermine the aura of precision and prestige that usually surrounds them. His mythologies become ours too, as his work requires viewer involvement: one rudimentary Montana landscape is elaborately titled The evil twister does its maiming and killing on the North Montana plains, but we have to supply the narrative. Complementing the paintings, brown-paper rock sculptures migrate around solo or group shows, exuding an air of futility. For the Ikon Gallery's off-site programme, Liversidge proposed filling Birmingham with inflatable carcasses, falconry displays and wolf howls emitted from speaker stacks in roving hire vans. (AF)

Born 1973 in Lincoln. Lives and works in London.
SELECTED RECENT EXHIBITIONS
2005   WHAT YOU'D EXPECT, Northern Gallery for Contemporary Art, Sunderland (touring)
        THE AMERICAN WEST, Compton Verney (curated by Jimmie Durham and Richard Hill)
2004   AIN'T NO LOVE IN THE HEART OF THE CITY, CBAT Gallery Cardiff
        THE WEST, Richard Salmon Gallery, London (solo)
        THROUGH THE WINDOW, Italian Cultural Institute, London

CAMILLA LØW

The work of Camilla Løw fuses surface, shape, colour and material to create sculptures that lean, float, dangle and balance. Their apparently sophisticated form and craftsmanship, which reference Scandinavian design and Russian Constructivism, belie their low-tech, hand-made origins. The artist's practice is straightforward in an old-fashioned, Op-Art way, but with a clear contemporary zip. Her most successful pieces combine economy of style with playfulness – suggesting they are about to fall over. Moving between Modernist purity and decorative device, they echo '60s Minimalist sculpture while invoking the more human aesthetic of mobiles or necklaces. Løw has inspired a number of students currently coming out of Scottish art schools and producing similar work, but she is arguably the best in her genre. Born in Oslo, she studied in Glasgow, and her sculptures have been presented recently in a number of solo and group exhibitions both in the UK and abroad. (PS)

Born 1976 in Oslo, Norway. Lives and works in Glasgow.
SELECTED RECENT EXHIBITIONS
2006   60 – SIXTY YEARS OF SCULPTURE IN THE ARTS COUNCIL COLLECTION, Yorkshire Sculpture Park
        Jack Hanley Gallery, San Francisco (solo)
2005   CONTEMPORARY NORDIC SCULPTURE 1980-2005, Wanas Foundation, Sweden
        Sutton Lane, London (solo)
2004   BRITANNIA WORKS, The British Council, Athens (curated by Katerina Gregos, cat.)

TIM MACHIN

Using everyday materials, Tim Machin creates intricate drawings and arrangements. Minute in scale, they are surprising and witty. Mountains (2003), for example, is constructed from a torn-off corner of a Guardian stock-market report; each graph is carefully sliced and folded to form a miniature mountain range. In Plastic Bag (2004), the artist reveals another hidden geographic feature by removing reindeer from a Christmas carrier bag, and in Untitled (Lone Stag in Landscape) (2004), he stands a paper-cut-out stag resolutely on a ball of Blu-tack. Machin's practice sits within a field of sculpture that displays a meticulous construction process, and demands an intense level of scrutiny. What distinguishes Machin from his peers, however, is his sheer enjoyment in the process of communicating beyond the double entendre of his material to touch on socio-economic conditions or the environment. Tim Machin was shortlisted for the Jerwood Drawing Prize in 2002. (CD)

Born 1978 in Sheffield. Lives and works in Manchester.
SELECTED RECENT EXHIBITIONS
2005   NEW WORK, LOT, Bristol
           THE OWL OF MINERVA, fa projects, London
           EMERGENCY2, Aspex Gallery, Portsmouth
2002   JERWOOD DRAWING PRIZE, Jerwood Space, London

HEATHER AND IVAN MORISON

Living and working together since 2001, Heather and Ivan Morison have created a diverse and prolific body of work. They've circulated news about their allotment on printed cards; they've sold their produce outside art institutions; they've written a science-fiction novel on board a cargo ship; they've produced films, slide shows and newspaper ads. They present vivid experiences in a deadpan way, with transmission playing a vital role; their works take shape between source and reception, ultimately residing in the receiver's mind. The artists' practice is also a cumulative portrait of their marriage, which has its own public identity ('Mr & Mrs Ivan Morison'). At the same time, they document others' enthusiasms; their route through Europe, Asia, and New Zealand (exhibited under the quasi-scientific title Global Survey) was suggested by people they met along the way. The Morisons balance the medium-specificity of Conceptual art and the exact specialisms of science with the true sensibility of the Victorian amateur, for whom knowledge has no borderlines. (AF)

Born 1973 in Desborough/1974 in Nottingham. Both live and work in Arthog, North Wales.
SELECTED RECENT EXHIBITIONS
2005   BRITISH ART SHOW 6, Baltic Gateshead, Manchester, Nottingham and Bristol
           STARMAKER, Charles H Scott Gallery, Vancouver
           SOUTH LONDON GALLERY DENDROLOGY FIELD TRIP AND TREE TOUR, South London Gallery
           ART SHEFFIELD 05, SPECTATOR T, Sheffield Contemporary Art Forum
           THE REAL IDEAL, Millennium Galleries, Sheffield

Cathay Pacific Boeing 747-467

JOHN MULLEN AND LEE O'CONNOR

Edinburgh College of Art graduates John Mullen and Lee O'Connor have recently begun working together in a way that revolves around their studio practices in general, and their shared studio in particular. Their video works and live performances are cheaply made and drunkenly orchestrated attempts to improve their working practices through dialogue. The artists' collaborative video works are always exhibited alongside their individual practices, which shift from foregrounding the videos and performances to providing a backdrop or a set of props for their seemingly more authentic actions. O'Connor develops his Scottish imagery using things he finds around him every day, like tissue paper, crayons and sugar paper. Mullen also works with scraps, bits of old cloth, discarded carpet, spray paint and toilet roll to produce works that have the allure of heroic early Modernism. In their own ways, they both constantly seek to eradicate the chic from the shabby. (NM)

Born 1972 in Edinburgh/1974 in Fort William. Both live and work in Edinburgh.

SELECTED RECENT EXHIBITIONS

2005   TEAMING, Embassy Gallery, Edinburgh

CAMPBELL'S SOUP, Glasgow International, Glasgow School of Art

2004   SHADOW CABINET, Magnifitat (Mullen)

2003   NEW WORK SCOTLAND, Collective Gallery, Edinburgh (O'Connor)

MILLIGAN'S ISLAND, The International3, Manchester (Mullen)

MORE BOOTS = MANY ROUTES, Transmission Glasgow (O'Connor)

KATIE ORTON

Katie Orton's practice is very strongly rooted in the surrealism of two antithetical figures: Jean Cocteau and Ivor Cutler. This odd coupling has so far produced sculptures and drawings with a clumsy beauty and a sentimental fringe that are the result of sharply skewed observation and documentation – Bet Lynch meets David Lynch. These qualities are particularly evident in the artist's scribbled books, which focus on failed attempts to gentrify Scottish society by paternalistic design. Hence we get the wet dream of sophistication (Art-Deco, catholic tastes, Cocteau) coupled with dryly observed nuggets drawn from neglected aspects of the everyday (Fablon-lined pool halls, Presbyterian iconoclasm, Ivor Cutler). Another main theme in Orton's practice is the forthcoming ban on smoking in Scotland; her own habit, and the role of the cigarette in our culture, leads her to examine the object that Oscar Wilde described as "the perfect type of a perfect pleasure. It is exquisite, and it leaves one unsatisfied." (NM)

Born 1980 in Glasgow. Lives and works in Edinburgh.

SELECTED RECENT EXHIBITIONS

2006   GREAT ARTSPECTATIONS, The Embassy Gallery, Edinburgh

2005   Zoo Art Fair, London (with The Embassy)

2004   THE SIX RIGHT ONES, Troy, Stuttgart

BRONTE GOES TO HOLLYWOOD, Wuthering Heights, Edinburgh

MICK PETER

Mick Peter's sculpture is highly improbable: funny, clever, vulgar, subtle, stupid, intelligent and breezily complicated. It may look improvised and slack, but it is considered and tight. The range of materials Peter employs is broad, unconventional and often messy. Concrete, gaffer tape, cardboard, porcelain, rubber, foam, polystyrene are combined to create objects that interweave familiarity and oddity. Ask Mick Peter for a list of sources and you'll get, among others things, Sol LeWitt, Zap magazine, American roadside sculpture, Joan Didion, prog rock, Alasdair Gray, and psychedelic posters. These sources are not, however, chosen simply for their recherché quality: they provide raw material for a dedicated investigation into visual symbolism, linguistic structures and iconographic efficiency. (Peter is a PhD candidate whose practice-led research explores these areas.) That which at first appears conceptually ludicrous and aesthetically irresponsible gradually tautens into focus, possibly revealing bits of brutalist architecture or clunky public sculptures. (JC)

Born 1974 in Berlin. Lives and works in Glasgow.
SELECTED RECENT EXHIBITIONS
2005   LIKE IT MATTERS, CCA: Centre for Contemporary Art, Glasgow
        Galerie Nomadenoase at Fortescue Avenue/Jonathan Viner, London (solo)
2004   DARLING TREES DO YOU NEED TELLING?, Switchspace, Glasgow
2003   VATHEK PRINT WORKS, Transmission Gallery, Glasgow
        NURSERY WORLD, Galerie Jennifer Flay, Paris

ALEX POLLARD

A major figure in contemporary Scottish art, Alex Pollard produces small sculptures and low-relief paintings out of plaster that he manipulates while it's drying. His subjects range from invented dinosaurs to human body parts (heads, smiley faces, arms, hands); some elements are hand-carved out of putty, then painted, creating an entire trompe l'oeil surface. Pollard's practice reflects his continued fascination with, and reaction against, our culture's obsession with what is real — he describes his works as 'hyper self-reflective', referring to themselves as both representations, and representations of art, and he goes on to remark that his chosen medium is 'chasing its own tail'. As well as his graphic output, Pollard is known for his satirical collaboration with Ian Hetherington in the form of Mainstream, a copied zine they have published together since the late '90s, thus carrying forward the honoured Scottish tradition of an independent and radical cultural press. (NM)

Born 1977 in Brighton. Lives and works in Glasgow.
SELECTED RECENT EXHIBITIONS
2006   The Approach, London (solo)
2005   SELECTIVE MEMORY, Scotland at The Venice Biennale,
        touring to The Gallery of Modern Art, Edinburgh
2004   SEE-THROUGH MASK, Sorcha Dallas (off-site project), Basel Art Fair Statement, Basel
        JOHN MOORES PAINTING PRIZE 2004, Walker Art Gallery, Liverpool

MARY REDMOND

Heavy theory may not sit easily with Mary Redmond's work, but this does not render it trivial or slight. From everyday rubbish, Redmond conjures gleaming bursts of beauty, moments of clarity rescued from obscurity by way of intuition. Her sculptures arise from a way of looking at the world that is neither heroically addressed to the future, nor critically consumed by the past. Instead they turn the forgotten and discarded detritus of the present into sparkling jewels. At the same time, her work is critically and historically connected, with a relationship to modern traditions that springs from Duchamp and Picasso through to Surrealism and beyond. Furthermore, it might be viewed within the context of traditional feminist issues (care, craft, decoration). Ultimately, however, it is extemporised, fresh and surprising — strongly linked to sensibilities derived from music and clubbing, fashion and street style. Mary Redmond's work may appear reticent and modest, but this should not lead to us to underestimate it. (JC)

Born 1973 in London. Lives and works in Glasgow.
SELECTED RECENT EXHIBITIONS
2005   JUNO AND THE STALLION, Talbot Rice Gallery, Edinburgh (solo)
2004   Galerie Christian Drantman, Brussels (solo)
        I FEEL MYSTERIOUS TODAY, Palm Beach Institute of Contemporary Art
2002   LADIES ROCK, The Changing Room, Stirling, Scotland (Scottish National Park Project
        with The Centre, Glasgow)

HANNAH RICKARDS

Nature and artifice are tightly woven in Hannah Rickards' sound art. Some of her works consist of her own voice reproducing sounds from the natural world. To simulate six common birds in Birdsong, she first slowed recordings of their songs, and altered the pitch so it was within her range. Then, after she recorded them, she changed the pitch back. This elaborate production machinery positions the artist's performance as a medium between the original sound and its re-production. Rickards is currently exploring the possibility of recreating the 'sounds' of the Northern Lights — sounds whose existence is denied by the scientific community. Here, as in so much of her practice, Rickards is interested in the electromagnetic transition of a visual experience into sound, a concern that builds on the research she carried out in her recent work Thunder (in which an orchestra reproduced the sound of a thunder clap). (FM)

Born 1979 in London. Lives and works in London.
SELECTED RECENT EXHIBITIONS & PERFORMANCES
2005-6 THUNDER, commissioned by Media Art Bath, Central United Reformed Church, Bath;
        68 Hope Street, Liverpool; performance at the South London Gallery (23.03.06)
2005   POSTERS SHOW, LOT, Broadmead, Bristol
2004   SECRET GARDEN: BEATING ABOUT THE BUSH, South London Gallery, London
2003   BLOOMBERG NEW CONTEMPORARIES 2003, Cornerhouse, Manchester and 14 Wharf Road, London
        BOOTLEG, Spitalfields Market, London

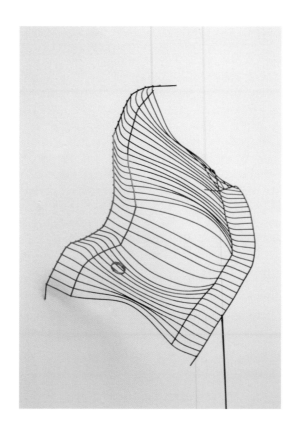

ANTHONY SHAPLAND

In one of his novels, John McGregor describes the moment when night turns into day: "It's the briefest of pauses, with not time enough to even turn full circle and look at all the lights this city throws out to the sky, and it's a pause which is easily broken. A slamming door, a car alarm, a thin drift of music from half a mile away, and already the city is moving on, already tomorrow is here." This is the territory inhabited by Anthony Shapland, well known for his direction of the G39 artspace in Cardiff, Shapland has recently displayed a mature capacity to find something remarkable in the mundane details of the nocturnal city. Rise (2003) captures the moment a street lamp switches on, warming into an orange glow. Nocturne (2004 – 05), a two-channel video installation that tracks night-time activity around a street doorway, convinces through its subtle use of the camera as an instrument of stillness and investigation. (CD)

Born 1971 in Pontypridd, South Wales. Lives and works in Cardiff.
SELECTED RECENT EXHIBITIONS
2005   SOMETHING OF THE NIGHT, Leeds City Gallery
        OVER AND OVER, AGAIN AND AGAIN, CAC, Vilnius, Lithuania
2004   UNPLUMBED, Keith Talent Gallery, London
2003   AFTERLIFE, VANE, Newcastle and Chapter, Cardiff

NOMINATED ARTISTS IMAGE CAPTIONS

P. 96 (top) Karla Black
Detail from THE DIFFERENTS (2005)
Wood, paint, earth, steroid ointment tubes,
Boots pharmacy paper bag
Courtesy the artist and Mary Mary, Glasgow

P. 96 (bottom) Simon Blackmore
WEATHER GUITAR (2005)
Flamenco guitar, MDF, metal, weather sensors,
step motors, electronic components, wires,
electricity. Installation view, IKON Gallery
Courtesy the artist

P. 98 (top) Simon and Tom Bloor
WHAT ARE THE SENSES? (2005)
Installation view, Lubetkin & Tecton Kiosk,
Dudley Zoological Gardens
Photocopies, plywood, screen & litho-
printed publication
Courtesy the artist

P. 98 (bottom) Matt Bryans
UNTITLED (2006)
Erased newspaper cutting, 60 x 41.6cm
Courtesy Kate MacGarry, London

P. 100 (top) Tomas Chaffe
SOLAR-PANELLED FLOOR AT THE ERNST
LUDWIG KIRCHNER MUSEUM (2005)
Digital print, A4
Courtesy the artist

P. 100 (bottom) Ruth Claxton
Detail from I THOUGHT I WAS THE
AUDIENCE AND THEN I LOOKED AT YOU (2005)
Ceramic, fimo, 20 x 5 x 5cm
Courtesy the artist

P. 102 (top) Luke Collins
Still from EPISODE 6 (2006)
DVD, 4min 30secs.
Courtesy the artist

P. 102 (bottom) William Daniels
THE FOREST ON ITS HEAD (2005)
Oil on board, 34 x 26.5cm
Courtesy Gaby & Wilhelm Schürmann, Herzogenrath

P. 104 (top) Kate Davis
PROPOSE I (2005)
Installation view, The Breeder, Athens
Silkscreen on cotton, ladder and acrylic paint
Courtesy Sorcha Dallas, Glasgow

P. 104 (bottom) Will Duke
Still from WE FASHIONED THE CITY
ON STOLEN MEMORIES (2005)
Computer generated animation, 6 mins.
Courtesy the artist

P. 106 (top) Marte Eknaes
YOUNG OFFENDER (2005)
Mirror Perspex (black), clear Perspex,
chain, metallic tap, 98 x 45 x 65cm
Courtesy the artist and Blum & Poe, Los Angeles

P. 106 (bottom) Erica Eyres
Still from KEY #02 (2005)
DVD, 6 mins.
Courtesy the artist

P. 108 (top) Laurence Figgis
THE GREAT MACGUFFIN (2005)
Ink, watercolour and crayon on paper,
82 x 353cm
Courtesy the artist

P. 108 (bottom) Josephine Flynn
GRADUATION (1998)
Colour photograph
Courtesy the artist

P. 110 (top) Rachel Goodyear
MEDICATE IN THE MISTAKEN BELIEF (2005)
Pencil on paper, 42 x 60cm
Courtesy the artist

P. 110 (bottom) Kate Owens and Tommy Grace
THE ILL TEMPERED WATERS (2005)
Performance/installation
Cooper Gallery, Duncan of Jordanstone
College of Art and Design, Dundee
Timber, lining paper, chipboard, 52 x 2 Litre
bottles of fizzy juice, 800 x 800 x 150 cm
Courtesy University of Dundee Exhibitions Dept.

P. 112 (top) Anthea Hamilton
UNTITLED (4-WAY) (2005)
Mixed media on board, 55 x 41.5cm
Courtesy the artist

P. 112 (bottom) Lucy Harrison
WAR GAMES (2005)
Vinyl on wall, 40 x 30cm
Courtesy the artist

P. 114 (top) Kathleen Herbert
Still from GRANDE SPAGNA (2005)
8-monitor DVD Installation
Dur. variable from 54 secs. to 3 min 45 secs.
Courtesy the artist

P. 114 (bottom) Dean Hughes
DRAWING AND HOLE PUNCHED A4 PAPER NO. 2 (2006)
Pins, thread and A4 paper
Courtesy Jack Hanley Gallery

P. 116 (top) i-cabin
Courtesy i-cabin

P. 116 (bottom) Torsten Lauschmann
COLD WATER QUARTET (2005)
Live at ICA, London
Conductor/loops, Single Fish – Bass,
Special Fish – Piano/Keyboards,
Animal – Drums/Percussion
Courtesy the artist

P. 118 (top) Peter Liversidge
THE LAST FRONTIER (2003 – 2005)
Mixed media
Dimensions variable
Courtesy the artist

P. 118 (bottom) Camilla Løw
Installation View, Sutton Lane (2004)
Courtesy Sutton Lane

P. 120 (top) Tim Machin
THE LAST PLANE ANYWHERE (2004)
9 x 14 cm
Courtesy the artist

P. 120 (bottom) Heather and Ivan Morison
FLOWERS WITH WOOD PIGEONS (2004)
Lambda print, 80 x 80cm
Courtesy Danielle Arnaud Contemporary Art

P. 122 (top) John Mullen & Lee O'Connor
WORKERS STATE 1 (2003)
Laser print, 42 x 29.7cm
Courtesy the artists

P. 122 (bottom) Katie Orton
TURN OR BURN (2005)
Acrylic paint on canvas
180 x 100cm approx.
Courtesy the artist

P. 124 (top) Mick Peter
NOPE (2005)
Ping pong balls, cement and polystyrene
147 x 200 x 74cm
Courtesy the artist

P. 124 (bottom) Alex Pollard
FIGURES (2005)
Acrylic paint, plaster and epoxy putty
23 x 24 x 3.5cm
Courtesy Sorcha Courtesy Dallas, Glasgow

P. 126 (top) Mary Redmond
DAQUARI (2005, detail)
Wood, metal, paint
180 x 43 x 16cm
Courtesy The Modern Institute

P. 126 (bottom) Hannah Rickards
Photograph taken during recording
of THUNDER (2005)
Sound installation, commissioned by Media Art Bath,
Central United Reformed Church, Bath
Courtesy the artist

P. 128 Anthony Shapland
Still from NIGHTWORKERS (2005)
DVD, 18 mins.
Commissioned by Vitrine
Courtesy the artist

CONTRIBUTORS
COLOPHON